Fundamentals of Inorganic and Organometallic Polymer Science

Fundamentals of Inorganic and Organometallic Polymer Science

By

Christian Agatemor

University of Miami, USA
Email: cxa903@miami.edu

Kajal Ghosal

Jadavpur University, India
Email: kajal.ghosal@gmail.com

Samuel Fura

University of Miami, USA
Email: sxf819@miami.edu

and

Peter J. S. Foot

Kingston University, London, UK
Email: p.j.foot@kingston.ac.uk

ROYAL SOCIETY
OF **CHEMISTRY**

Print ISBN: 978-1-78801-590-5
PDF ISBN: 978-1-83767-232-5
EPUB ISBN: 978-1-83916-322-7

A catalogue record for this book is available from the British Library

The Royal Society of Chemistry is a charity, registered in England and Wales, Number 207890, and a company incorporated in England by Royal Charter (Registered No. RC000524), registered office: Burlington House, Piccadilly, London W1J 0BA, UK, Telephone: +44 (0)20 7437 8656.

Visit our website at books.rsc.org

Preface

This monograph describes the basic concepts, syntheses, properties, characterisation, and applications of inorganic and organometallic polymers. These polymers incorporate inorganic elements or transition metals as building blocks. The presence of these building blocks impacts unique properties uncommon in organic polymers and is potentially useful for designing many functional materials.

This monograph is intended for lecturers teaching a one-semester advanced polymer science course and for self-study researchers working with polymers. The first chapter introduces the core concepts of polymer science, including the types of polymerisations and polymer nomenclature. The chapter also discusses why inorganic and organometallic polymers are better suited than organic polymers in certain applications. Subsequent chapters walk the reader through an in-depth discussion of the syntheses (Chapters 2 and 3), characterisation (Chapter 4), properties (Chapter 5), examples (Chapter 6), and applications (Chapter 7) of inorganic and organometallic polymers. In each chapter, worked examples and review questions are included to assist readers in assessing their understanding of the concepts.

We intend this monograph to be a helpful classroom text for students and lecturers and a self-educating introduction to inorganic and organometallic polymers for polymer scientists and engineers. Additionally, when used with the journal references listed at the end of each chapter, the monograph provides the reader with a thorough and expert grounding in the field of inorganic and organometallic polymers.

Christian Agatemor, Kajal Ghosal, Samuel Fura
and Peter J. S. Foot

Fundamentals of Inorganic and Organometallic Polymer Science
By Christian Agatemor, Kajal Ghosal, Samuel Fura and Peter J. S. Foot
© Christian Agatemor, Kajal Ghosal, Samuel Fura and Peter J. S. Foot 2024
Published by the Royal Society of Chemistry, www.rsc.org

Contents

Fundamentals of Inorganic and Organometallic Polymer Science
By Christian Agatemor, Kajal Ghosal, Samuel Fura and Peter J. S. Foot
© Christian Agatemor, Kajal Ghosal, Samuel Fura and Peter J. S. Foot 2024
Published by the Royal Society of Chemistry, www.rsc.org

1 Concepts in Polymer Chemistry

1.1 Introduction

Human civilisation relies greatly on developing and accessing materials, which include ceramics, metals, and polymers. Indeed, archaeologists classify the timeline of human technological prehistory into the Stone Age, Bronze Age, and Iron Age based on the materials that were predominantly used in designing the tools of that era. Therefore, it is fair to assert that we currently live in the Polymer Age because most contemporary technologies use polymers – plastics, fibres, elastomers, coatings, and adhesives. Polymer-based materials decorate our homes and offices, and we use them in almost every facet of life. Commodity materials such as household cooking utensils and clothing; engineering materials such as automobile and aircraft parts; and specialty materials such as medical implants, fibre optic systems, and even therapeutic and diagnostic agents are composed of polymers (Box 1.1). While the availability of inexpensive petroleum contributes to the revolutionary transition from metals and ceramics to polymers, advances in chemistry, which have enriched our understanding of polymers, undoubtedly contributed the most to the innovative use of polymers in improving our quality of life. Despite the achievements, chemists continue to push the frontiers of what is possible in the world of polymers. As an

---

Box 1.1 Learning outcomes.

By the end of this chapter, the student should:

1. Define and identify a polymer.
2. Differentiate organic from inorganic and organometallic polymers.
3. Classify polymers as inorganic, organometallic, or coordination.
4. Describe a monomer and a repeat unit.
5. Define polymer molecular weight and dispersity.
6. Differentiate an elastomer, a fibre and a plastic.
7. Classify polymers according to their structural and functional properties.
8. Name polymers according to IUPAC recommendations.

Fundamentals of Inorganic and Organometallic Polymer Science
By Christian Agatemor, Kajal Ghosal, Samuel Fura and Peter J. S. Foot
© Christian Agatemor, Kajal Ghosal, Samuel Fura and Peter J. S. Foot 2024
Published by the Royal Society of Chemistry, www.rsc.org

example, petroleum, the prime raw material for producing organic polymers, is becoming depleted, and most synthetic organic polymers are now considered pollutants due to their persistence in the environment. An attractive solution is to investigate and design inorganic and organometallic polymers, which offer the opportunity to integrate the electronic, redox, magnetic, and optical properties of inorganic elements and transition metals with the intrinsic structural properties of organic polymers. Therefore, the study of inorganic and organometallic polymers has far-reaching implications for sustaining the positive impact of polymers in our society (Box 1.2).

Box 1.2   Classes of contemporary synthetic polymers.

1. *Commodity polymers*: these polymers are precursors to bulk and high-volume materials used in our daily life. These materials range from low-cost items such as plastic bags, plates, and cups to high-value goods such as designer cloth and shoes (Figure 1.1). Examples of commodity polymers are polyethylene, polypropylene, poly(methyl methacrylate), polystyrene, and poly(vinyl chloride).

Figure 1.1   Examples of materials produced from commodity polymers including rain boots, bowls, chairs and kitchen utensils. Images courtesy of Pixabay.

2. *Engineering polymers*: these polymers feature thermal and mechanical properties such as strength, toughness, and melting, and glass transition temperatures superior to those of commodity polymers. The quest for these types of

Concepts in Polymer Chemistry

materials contributed to the interest in inorganics and organometallics. Engineering polymers are precursors to structural materials, including automobile parts, aircraft parts, gears, bearings, and electrical devices (Figure 1.2). Typical engineering polymers include nylon-6,6, polyether ether ketone, polytetrafluoroethylene, polycarbonate, and polybutylene terephthalate.

Figure 1.2 Examples of materials produced from engineering polymers, including computer mouse, nylon rope, car dashboard and telephone. Images courtesy of Pixabay.

3. *Specialty polymers*: these are low-volume polymers specifically designed for an application. They possess a unique functional group or structure that imparts the desired property. The design of inorganic and organometallic polymers represents a strategy in imparting unique functional properties to polymers. Specialty polymers are used in the production of materials for the biomedical, information and energy industries. In biomedicine, for example, specialty polymers are used as therapeutics, biosensors, drug delivery vectors, and implants (Figure 1.3). At the same time, in the energy industry, they are fabricated into conducting polymers and solar energy harvesting materials. Examples of these polymers are polysiloxanes, poly(lactic acid-*co*-glycolic acid), immunoglobulin, hyaluronic acid, polyamidoamine dendrimers, polypyrrole, polythiophene, and polyphenylene vinylene.

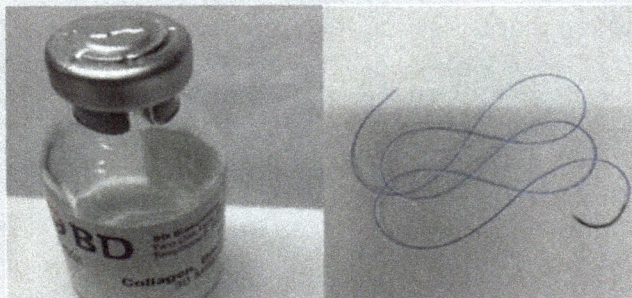

Figure 1.3 Examples of materials produced from specialty polymers. Collagen (in the vial on the left) is a protein used in many biomedical applications, while the suture (on the right) is derived from medical-grade polypropylene.

1.2 Historical Development

In 1907, Leo Baekeland invented Bakelite, the first successfully commercialised synthetic polymer. This invention contributed immensely to the development of polymer science. Other polymers such as polysiloxanes, celluloid, and vulcanised rubber predated this invention (Box 1.3), but the exceptional electrical, heat and chemical

---

Box 1.3   Timeline in polymer science.

Modern polymer science was inspired by the quest for surrogates of raw materials such as fibre, ivory, and natural rubber or the need to improve the properties of these raw materials. Below is a timeline of the development of polymers.

1823   C. Macintosh invented a process to use natural rubber to make waterproof fabrics. These waterproof garments were stiff in winter and sticky in summer due to the low glass and melt transition temperatures of natural rubber.

1833   J. J. Berzelius coined the term "polymer" from the Greek words "poly" meaning "many" and "meros" meaning "part". His original intent was to describe compounds of the same chemical composition and empirical formulas but of different molecular formulas and properties. This original definition considers benzene (C_6H_6) a polymer of acetylene (C_2H_2)! Nevertheless, in keeping with the original intent of the compositional relationship between different compounds and the relevance of the Greek meaning, modern chemists use the term to describe macromolecules obtained by linking many (poly) low molecular weight compounds (meros).

1833   H. Braconnot described the nitration of cellulose to nitrocellulose, laying the foundation for synthesising the first thermosetting plastic.

1835   H. V. Regnault synthesised poly(vinyl chloride) (PVC), but the polymer was patented in 1912 by F. H. A. Klatte, who used sunlight to initiate the polymerisation.

1839   C. Goodyear discovered sulphur-based vulcanisation of natural rubber. He obtained the patent in 1844. Vulcanisation converts tacky rubber to an elastomer, toughens natural rubber and increases its glass and melt transition temperatures, leading to many rubberised materials, including waterproof garments. Vulcanised natural rubber was the first commercially successful product of polymer research and gave birth to the rubber industry.

1845   C. F. Schönbein improved the nitration of cellulose, creating a soluble nitrocellulose derivative.

1852   C. Löwig synthesised polystannane, an organometallic polymer.

1856   A. Parkes obtained a patent to produce the first thermoplastic, celluloid, from nitrocellulose. Celluloid, which was also called synthetic ivory, was used to make many plastic materials such as combs, toys, and billiard balls.

1861   T. Graham advanced the aggregation theory to explain the molecular structure of polymers, proposing that polymers are aggregates of small molecules held together by unknown intermolecular forces.

1880   G. W. A. Kahlbaum synthesised poly(methyl acrylate); however, its commercial production started only in 1927.

1884   H. B. Comte de Chardonnet patented a method to produce artificial silk, later known as rayon, from nitrocellulose.

1895   H. N. Stokes reported the synthesis of the synthetic inorganic polymer polydichlorophosphazene (Figure 1.4a).

1901   F. S. Kipping coined the terms "silicoketone" and "silicone" to describe polydiphenylsiloxane (Figure 1.4c). His pioneering work in organosilicon chemistry led to the development of polysiloxanes, which are among the most important classes of inorganic polymers.

1907   L. Baekeland synthesised a thermosetting phenol–formaldehyde resin, called Bakelite, which became the first synthetic polymer with commercial success.

1920   H. Staudinger proposed that polymers are an extended chain of short repeat units of small molecules linked by covalent bonds, repudiating the aggregation theory of T. Graham.

1921   F. S. Kipping synthesised poly(diphenylsilane), a polysilane (Figure 1.4d), another class of inorganic polymers.

1935   W. H. Carothers synthesised nylon-6,6, the first nylon, at the DuPont Experimental Station.

1955   G. Natta applied the Ziegler–Natta catalyst to synthesise the first stereoregular polyolefins.

1955   F. S. Arimoto synthesised poly(vinyl ferrocene) homo- and copolymers (Figure 1.4b), archetypal organometallic polymers.

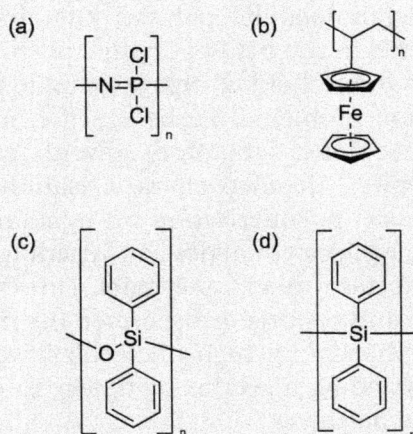

Figure 1.4   Repeat units of historical inorganic and organometallic polymers: (a) polydichlorophosphazene, (b) poly(vinyl ferrocene), (c) polydiphenylsiloxane, and (d) polydiphenylsilane.

resistances of Bakelite proved extremely useful in the emerging electrical and auto-mobile industries of the time. Despite the increasing commercialisation of polymers at that time, a clear understanding of their structure and the molecular forces behind their extraordinary properties remained elusive. Thomas Graham, who was dubbed the father of colloid chemistry, proposed in 1861 that polymers are colloidal aggregates of small molecules held together by mysterious intermolecular interactions (Box 1.3). This postulation persisted through the late 19th century and early 20th century and enjoyed the support of other leading chemists of the time, including Nobel laureate Emil Fischer, who posited that the measured high molecular weights of polymers were apparent values resulting from aggregated small molecules.

In 1920, Hermann Staudinger proposed that polymers were an extended chain of atoms built from small molecule-derived repeat units held together by covalent bonds. Staudinger's hypothesis repudiated Graham's aggregation theory and earned him the Nobel Prize in Chemistry in 1953. The 1930s saw further developments, including the work of Wallace Hume Carothers, that experimentally validated Staudinger's hypothesis and led to the birth of experimental polymer science and the commercialisation of more synthetic polymers, such as neoprene and polyamide (nylon) fibres. The invention of the Ziegler catalyst in 1952 by Karl Ziegler and its application by Giulio Natta in the polymerisation of α-olefins to stereoregular polyolefins having mechanical properties that are superior to those of non-stereoregular polyolefins revolutionised the polymer industry and led to the large-scale commercialisation of polymers.

1.2.1 Development of Organometallic Polymers

Although the developments mentioned above focus on organic polymers, concurrent advances in organometallic chemistry, particularly the discovery of ferrocene in the 1950s, triggered interest in the possibilities of organometallic polymers. In 1955, for example, Arimoto and Haven reported the first polymerisation of a vinyl derivative of ferrocene.[1] By 1966, Rosenberg and Neuse employed vinyl ferrocene to synthesise a well-characterised, linear organometallic polymer that had a molecular weight of 7000 Da.[2] Unlike developments in the field of organic polymers that snowballed in the early 20th century, advances in the field of organometallic polymers were slow due to synthesis and characterisation problems. To be specific, most of the organometallic polymers were insoluble in common laboratory solvents precluding the synthesis of high molecular weight polymers, detailed characterisation, and processing into finished products. Simultaneously, polymerisation of organometallic monomers, except vinyl-containing monomers, utilises condensation reactions (discussed in Chapter 2) that generally yield low molecular weight polymers. However, the early 1990s saw a breakthrough when Manners and coworkers discovered the ring-opening polymerisation of strained [1]silaferrocenophanes to high molecular weight polyferrocenylsilanes (discussed in Chapter 3),[3] providing a synthetic strategy to design a class of high molecular weight organometallic polymers.

Before the synthesis of polyferrocenylsilanes, inorganic polymers such as poly-siloxanes, polysilanes, and polyphosphazenes (discussed in Chapter 7) had been successfully synthesised (Box 1.3). As an example, the synthesis of polysiloxane, a commercially available inorganic polymer, dramatically blossomed after Kipping and Lloyd used the Grignard process to synthesise polydiphenylsiloxane in 1901.[4] Advances in later years in the field of synthetic and analytical chemistry led to the synthesis and characterisation of different types of inorganic and organometallic polymers that will be elaborated on in later chapters. As a prelude to these chapters, the rest of this chapter will discuss some basic principles and concepts related to polymer science.

1.3 Polymers, Oligomers, and Monomers

Polymers and oligomers are high molecular weight molecules built by linking several repeat units derived from low molecular weight molecules called monomers (Table 1.1). The number of repeat units, also known as "mer," distinguishes polymers from oligomers. Generally, oligomers consist of fewer repeat units than polymers, but since this distinction is blurry (there is no consensus on the number of repeat units that define an oligomer), we will describe both macromolecules as polymers throughout this monograph. The elemental composition of the monomers determines polymers, with those composed solely of organic groups being termed organic polymers (Box 1.4). At the same time, polymers that lack organic groups are known as inorganic polymers. In contrast, hybrids, comprising organic and inorganic groups, are termed organic–inorganic hybrid, organometallic or coordination polymers (Box 1.4).

   The properties of inorganic and organometallic polymers differ from those of their organic counterparts. Organometallic polymers, for instance, could exhibit the properties of transition metals such as optical properties, redox activity, photoactivity, and bioactivity (discussed in Chapter 5) that are uncommon in their organic counterparts.

Table 1.1 Structural differences between polymers, monomers, and repeat units.

Box 1.4 Organic, inorganic, organometallic, *vs.* coordination polymers (Scheme 1.1).

Organic polymers: polymers with a skeletal structure composed only of carbon atoms that are covalently bonded to hydrogen, oxygen, nitrogen, sulphur, phosphorus, or halogen atoms. Examples include polyolefins, polystyrene, polyacrylates, poly-amides, polyesters, polytetrafluoroethylene, polyphenylene sulphide, *etc.*

   Inorganic polymers: polymers with a skeletal structure that lacks carbon atoms. Examples include polysilazanes, polysulphides, polysilicates, polysiloxanes, poly-silanes, polyphosphazenes, *etc.*

   Organic–inorganic hybrids: polymers with a skeletal structure composed of organic and inorganic moieties. These polymers should not be confused with co-ordination and organometallic polymers, which contain metal atoms. Examples include polyorganophosphazenes, polyorganosilanes, *etc.*

   Coordination polymers: polymers with a skeletal structure that contains organic ligands that are coordinatively bonded to metal atoms. These polymers include metal–organic frameworks, metal-complexed polypyridines, polyporphyrins, poly-carboxylates, *etc.*

   Organometallic polymers: polymers with a skeletal structure that includes carbon atoms that are covalently bonded to metal atoms. Examples include poly(vinyl me-tallocene), polystannane, polygermanes, *etc.*

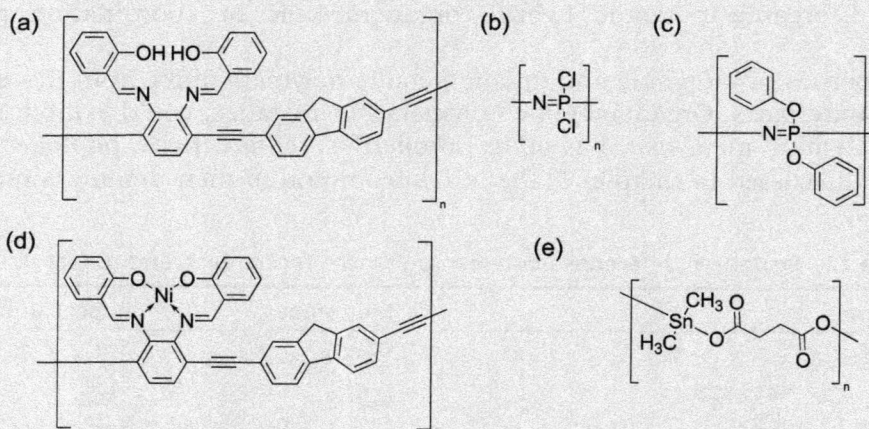

Scheme 1.1 Schematics showing the skeletal structure of repeat units in (a) organic, (b) inorganic, (c) organic–inorganic hybrid, (d) coordination, and (e) organo-metallic polymers.

The presence of these properties informs the continuing exploration of these polymers as antifoulants, bactericides, photosensitisers, conducting polymers, and bioimaging probes (discussed in Chapter 6). However, most inorganic and organometallic polymers remain laboratory curiosity because of their intractability to current polymer processing methods, but a few inorganic polymers such as polysiloxane and polyphosphazene (discussed in Chapter 7) are in the market.

Box 1.5 Worked example 1.1.

Identify the monomer and repeat unit in the polymers below.

a.

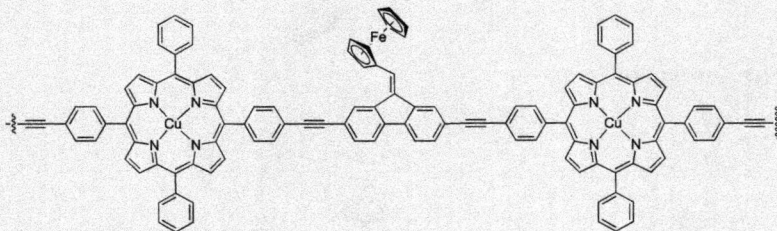

b.

c.

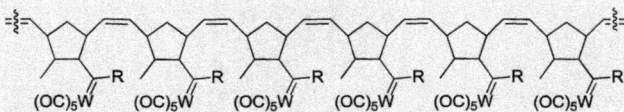

Solution:

a. The polymer is obtained *via* polycondensation of the monomer

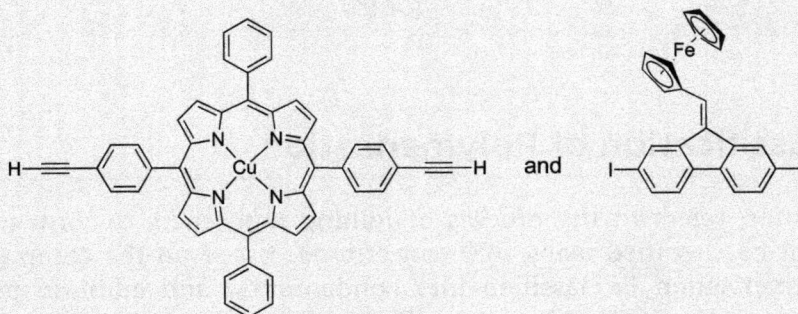

and the repeat unit is

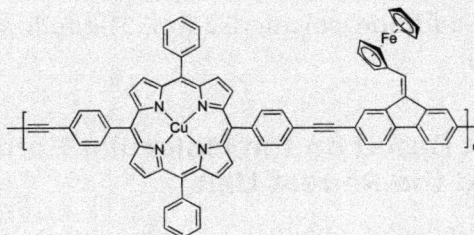

b. The polymer is obtained *via* ring-opening polymerisation of the monomer

and the repeat unit is

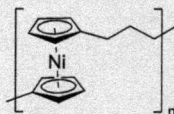

c. The polymer is obtained *via* ring-opening metathesis polymerisation of the monomer

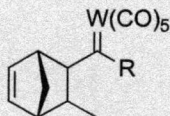

and the repeat unit is

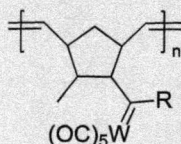

1.4   Classification of Polymerisation

Polymerisation refers to the process of joining monomers to form polymers. This process can be classified using different criteria. Based on the composition of polymers, polymerisation is classified into condensation and addition polymerisation, while based on the mechanism it is classified into step and chain polymerisation (discussed in Chapters 2 and 3). These classifications can be ambiguous for beginning students because some literature often substitutes step with condensation polymerisation and chain with addition polymerisation. The following subsections clarify the different classifications.

1.4.1   Classification Based on the Molecular Formulas of the Monomer and the Repeat Unit

W. H. Carothers, a pioneer of polymer science who synthesised the first nylon at DuPont in 1935 (Box 1.3), was the first to note the relationship between the molecular

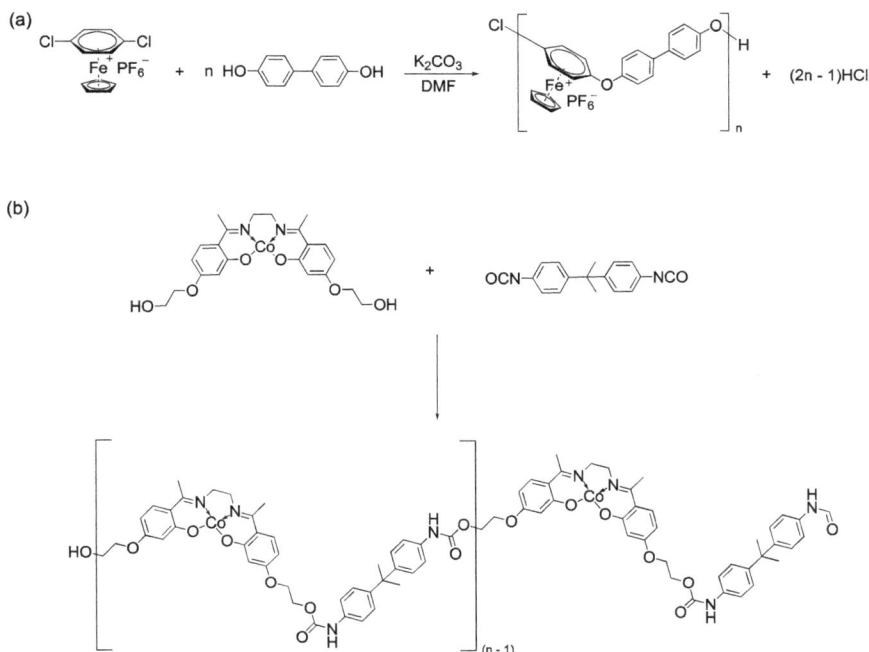

Figure 1.5 Examples of condensation polymerisation, in which the molecular formula of the repeat unit differs from that of the monomer. The polymerisation occurs with the elimination of a small molecule (a) or results in polymers containing repeat units linked by a functional group (b).

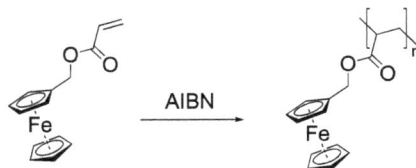

Figure 1.6 An example of addition polymerisation that occurs without eliminating small molecules and in which functional groups do not connect the repeat units.

formulas of the monomers and the repeat unit in a polymer. He used this relationship to distinguish condensation from addition polymerisation. Carothers described condensation polymerisation as that in which the molecular formula of the monomer differs from that of the polymer repeat unit (Figure 1.5).[4] In contrast, the molecular formula of the monomer is identical to that of the repeat unit in a polymer derived from addition polymerisation (Figure 1.6).[4] He conceptualised polymerisation as the coupling of monomers to form polymers with or without the elimination of small molecules. This classification, however, is ambiguous because some universally accepted condensation polymerisations do not result in the elimination of small molecules (Figure 1.5b). This ambiguity notwithstanding, a polymerisation is generally classified as condensation or addition if it meets the requirements specified in Box 1.6.

> **Box 1.6**   Condensation *vs.* addition polymerisation.
>
> Carothers distinguished between condensation and addition polymerisations based on the structural relationship between the monomer and repeat unit in a polymer. The key features used to classify these polymerisations are given below.
>
>   A condensation polymerisation meets one of these requirements:
>
> 1. Its synthesis involves the elimination of a small molecule such as H_2O, HCl, and NH_3.
> 2. The repeat unit of the polymer lacks some atoms that are present in the monomer from which it is derived or the product it forms after degradation.
> 3. Functional groups connect the repeat units.
>
>   A polymerisation is classified as an addition polymerisation if it meets one of the following requirements:
>
> 1. Occurs without the elimination of small molecules.
> 2. Functional groups do not link the repeat units.
> 3. The repeat units have the same composition as the monomer.

1.4.2   Classification Based on the Mechanism

The ambiguity inherent in classifying polymerisation based on differences between the molecular formulas of the monomer and repeat unit led to the use of the reaction mechanism as another framework for classification. Nobel laureate Paul J. Flory, a coworker of Carothers, reasoned that the prime difference between condensation and addition polymerisation lies not in the difference between the molecular formulas of the monomer and repeat unit but in their reaction mechanisms.[5] Flory showed that condensation polymerisation occurs *via* a stepwise intermolecular condensation reaction between complementary functional groups, while addition polymerisation proceeds through a chain mechanism that involves reactive centres.[5] His work, therefore, introduced the popular concepts of step and chain polymerisations (discussed in Chapters 2 and 3). Two key features – the nature of the reactive species and the degree of monomer conversion – distinguish step from chain polymerisation.

   In step polymerisation (discussed in Chapter 2), the polymer grows through a stepwise, random reaction between monomers to form dimers, trimer, tetramers, and other polymeric species that equally react among themselves to yield high molecular weight polymers (Figure 1.7a). The molecular weight of the polymer depends on the degree of monomer conversion. For instance, high molecular weight polymers are obtained only towards the end ($>98\%$ monomer conversion) of the polymerisation, meaning that only small- and intermediate-sized polymers exist in the reaction vessel at low monomer conversion (Figure 1.7c). In a chain polymerisation (discussed in Chapter 3), an initiator generates a radical, a cation, or an anion, which bears a reactive centre that sustains the polymerisation by successive addition of monomers to the reactive end of the growing polymer (Figure 1.7b). Chain polymerisation proceeds through initiation, propagation, and termination steps, yielding high molecular weight polymers at all degrees of monomer conversion, 0.1 or 98% (Figure 1.7b and c). The key features of step and chain polymerisations are given in Box 1.7.

(a)

Monomer	+	Monomer	⟶	Dimer
Dimer	+	Monomer	⟶	Trimer
Trimer	+	Monomer	⟶	Tetramer
Dimer	+	Dimer	⟶	Tetramer
Tetramer	+	Monomer	⟶	Pentamer
Trimer	+	Dimer	⟶	Pentamer
Pentamer	+	Monomer	⟶	Hexamer
Trimer	+	Trimer	⟶	Hexamer
x-mer	+	y-mer	⟶	(x+y)-mer

(b)

Initiator	⟶	Reactive species	Initiation
Reactive species	$\xrightarrow{\text{Monomer}}$	Monomer bearing reactive centre	Propagation
Monomer bearing reactive centre	$\xrightarrow{\text{Monomers}}$	Polymer bearing reactive centre	Propagation
Polymer bearing reactive centre	$\xrightarrow{\text{Polymer bearing reactive centre}}$	Polymer	Termination
Polymer bearing reactive centre	$\xrightarrow{\text{- H}}$	Polymer	Termination

(c)

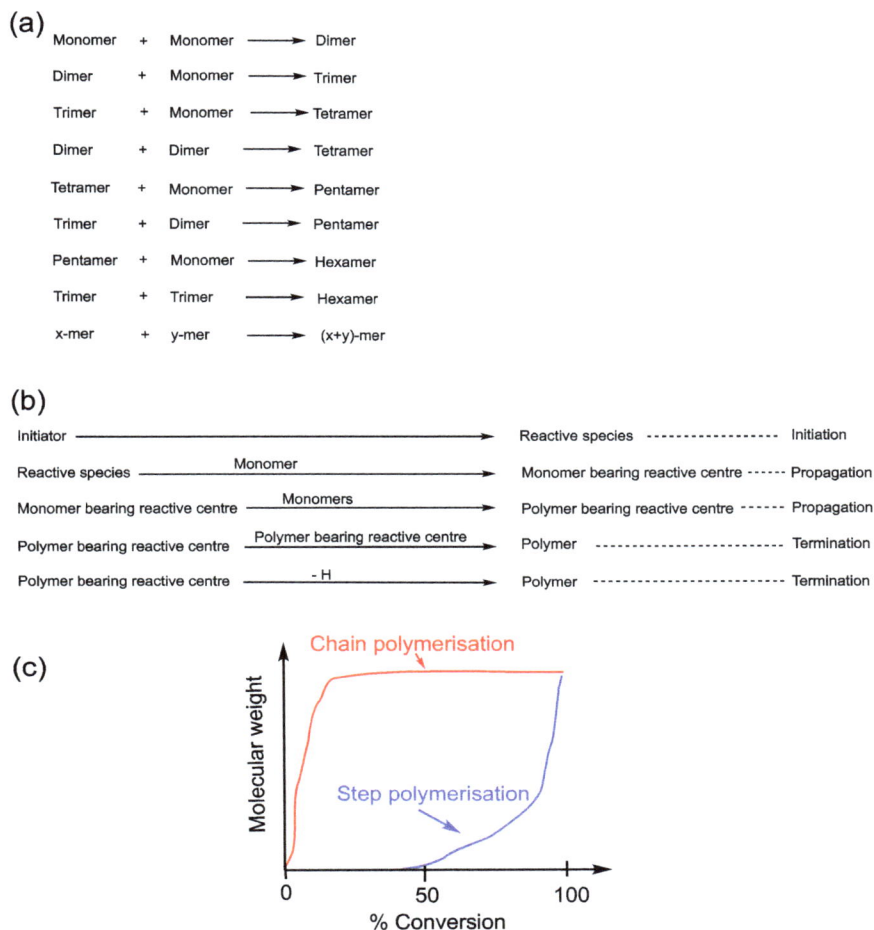

Figure 1.7 Schematics showing the mechanistic steps in (a) step and (b) chain polymerisation. (c) Relationship between the growth of molecular weight and the degree of monomer conversion.

Box 1.7 Step *vs.* chain polymerisation.

Step and chain polymerisations are classified based on the mechanism. The key mechanistic features of these polymerisations are given below.

In step polymerisation,

1. Monomers react with other monomers or polymeric species that bear a complementary functional group.
2. The molecular weight of the polymer depends on the degree of monomer conversion.

In chain polymerisation,

1. A monomer cannot react with another monomer or polymeric species unless they are radicals, cations, or anions.
2. The molecular weight of the polymer is independent of the degree of monomer conversion.

The above classifications should not be oversimplified because both mechanisms can operate in each polymerisation, leading to ambiguity. For instance, ring-opening polymerisation (ROP) (discussed in Chapter 3) is classified as a chain process because it comprises initiation, propagation, and termination steps. However, the propagation rate constants of ROP are several orders of magnitude lower than those of chain polymerisation and similar to those of step polymerisation. Because ROP can yield high molecular weight polymers, it is an alternative to step polymerisation that typically produces low molecular weight polymers. Also, in ROP, the dependence of polymer molecular weight on monomer conversion deviates from the characteristic behaviour of chain polymerisation. Instead, many ROPs follow the living polymerisation mechanism, where the termination step is suppressed and the molecular weight of the polymer increases linearly with the monomer to initiator ratio and monomer conversion (Box 1.8).

Box 1.8   Worked example 1.2.

Classify the following polymerisations based on the relationship between the monomer and repeat unit (condensation *vs.* addition polymerisation) and mechanism (step *vs.* chain polymerisation).

a.

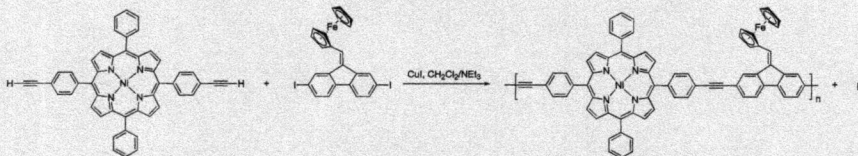

b.

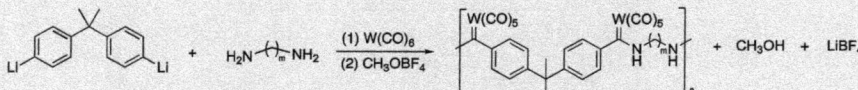

c.

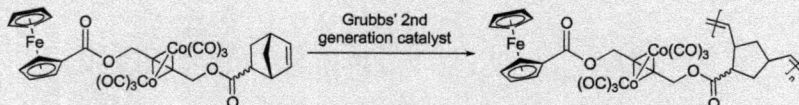

d.

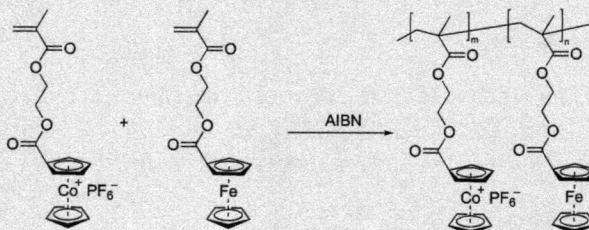

Solution:

(a and b) These are condensation polymerisations because there is the elimination of a small molecule and the molecular formula of the repeat unit differs from that of the monomer. Based on the mechanism, these are step polymerisations because they involve a reaction between monomers bearing complementary functional groups.

(c and d) These are addition polymerisations because there is no elimination of a small molecule. Based on the mechanism, these are chain polymerisations that involve an initiator.

In summary, to unambiguously classify a polymerisation, it is essential to describe all the characteristics – the products of the reaction, the relationship between the monomer and the repeat unit, and the polymerisation mechanisms (Box 1.9).

Box 1.9 Worked example 1.3.

An organometallic polymer consists of four fractions of different-sized macro-molecules, as shown in the table below. Calculate the number-average (M_n) and weight-average (M_w) molecular weights. Also, compute the dispersity.

Fraction	Number of moles	Molecular weight ($g\,mol^{-1}$)
A	3	72 000
B	9	115 000
C	7	68 000
D	5	95 000

Solution
Number-average Molecular Weight
Recall eqn (1.1):

$$M_n = \frac{\sum N_i M_i}{\sum N_i}$$

where N_i is the number of moles of polymer molecules having a molecular weight of M_i.

Then M_n is

$$\frac{\left(3 \text{ mol} \times 72\,000 \text{ g}\,mol^{-1}\right) + \left(9 \text{ mol} \times 115\,000 \text{ g}\,mol^{-1}\right) + \left(7 \text{ mol} \times 68\,000 \text{ g}\,mol^{-1}\right) + \left(5 \times 95\,000 \text{ g}\,mol^{-1}\right)}{3 \text{ mol} + 9 \text{ mol} + 7 \text{ mol} + 5 \text{ mol}}$$

$$M_n = 91\,750 \text{ g}\,mol^{-1}$$

Weight-average Molecular Weight
Recall eqn (1.3):

$$M_w = \frac{\sum W_i M_i}{\sum W_i} = \frac{\sum N_i M_i^2}{\sum N_i M_i}$$

where N_i is the number of moles of polymer molecules having a molecular weight of M_i.

Then M_w is

$$\frac{3 \text{ mol } \left(72\,000 \text{ g mol}^{-1}\right)^2 + 9 \text{ mol } \left(115\,000 \text{ g mol}^{-1}\right)^2}{3 \text{ mol } \left(72\,000 \text{ g mol}^{-1}\right) + 9 \text{ mol } \left(115\,000 \text{ g mol}^{-1}\right)}$$
$$\frac{+ 7 \text{ mol } \left(68\,000 \text{ g mol}^{-1}\right)^2 + 5 \text{ mol } \left(95\,000 \text{ g mol}^{-1}\right)^2}{+ 7 \text{ mol } \left(68\,000 \text{ g mol}^{-1}\right) + 5 \text{ mol } \left(95\,000 \text{ g mol}^{-1}\right)}$$

$$M_w = 96\,308 \text{ g mol}^{-1}$$

Dispersity

$$Đ = \frac{M_w}{M_n} = \frac{96\,308 \text{ g mol}^{-1}}{91\,750 \text{ g mol}^{-1}}$$
$$Đ = 1.05$$

1.5   Molecular Weights of Polymers

The molecular weight of a polymer is a fundamental property that influences the other properties, such as tensile strength and viscosity, which ultimately impact its processability and end-use. The relationship between processability and application means that the molecular weight must be optimised and well-characterised. As the molecular weight increases, intermolecular forces increase, leading to increased tensile strength and viscosity but decreased processability. The concept of molecular weight in polymer chemistry differs from its classical meaning in general chemistry. In general chemistry, a sample of a small molecule compound, such as sodium chloride, is understood to consist of molecular species of the same molecular weight. However, in polymer chemistry, a sample of a synthetic polymer consists of macro-molecules of different chain lengths, and therefore different molecular weights. Thus, the molecular weight of polymers is characterised by a wide statistical distribution that is rarely symmetrical, resulting in a molecular weight dispersity that originates from the statistical variations in the polymerisation conditions. The reported molecular weight of polymers is an average of the molecular weights of the large-, intermediate-, and small-sized macromolecular species present in the sample. To fully characterise a polymer, it is crucial to define this average and dispersity ($Đ$), which is the distribution of the different molecular weights within the polymer (discussed in Chapter 4).

The value of the determined average molecular weight depends on the experimental techniques used for the analysis. Techniques that quantify polymer end-groups, such as one-dimensional proton nuclear magnetic resonance spectroscopy (^{1}H NMR), colourimetry and potentiometric titration, or measurement of colligative properties, such as osmotic pressure, boiling point elevation, freezing point depression, and vapour pressure depression, report number-average molecular weight (M_n) (discussed in Chapter 4). The molecular weight is called number-average molecular weight because these techniques count the number of polymer molecules of a specific molecular weight in the sample. Mathematically, M_n is defined as the total weight of the polymer divided by the total number of polymers:

$$M_n = \frac{\sum N_i M_i}{\sum N_i} \tag{1.1}$$

Eqn (1.1) can also be expressed as

$$M_n = \sum x_i M_i \tag{1.2}$$

where the summation is over all of the different sizes of polymer molecules from $i = 1$ to $i = \infty$, N_i is the number of moles of polymer molecules having a molecular weight of M_i and x_i is the mole fraction of each polymer having a molecular weight of M_i.

On the other hand, techniques such as light scattering, ultracentrifugation, and diffusion-ordered NMR report weight-average molecular weight (M_w) based on polymer size (discussed in Chapter 4). Unlike M_n, M_w considers the weight of each polymer molecule with a specific molecular weight and is defined mathematically as

$$M_w = \frac{\sum W_i M_i}{\sum W_i} = \frac{\sum N_i M_i^2}{\sum N_i M_i} \tag{1.3}$$

Eqn (1.3) can also be expressed as

$$M_w = \sum w_i M_i \tag{1.4}$$

where the summation is over all the different sizes of polymer molecules from $i = 1$ to $i = \infty$, W_i is the weight of polymer molecules with a molecular weight of M_i, and w_i is the weight fraction of each polymer molecule with a molecular weight of M_i.

The M_w is biased towards large-sized macromolecules, while the M_n favours small-sized macromolecules. Because polymer properties depend more on the large-sized macromolecules, M_w is a better predictor of the behaviour of polymers than M_n. However, to compute dispersity ($Ð$), one must determine both M_w and M_n.

M_w is always greater than M_n (Figure 1.8a) unless all polymer molecules are of the same weight, which is most unlikely due to the statistical variations that exist during polymerisation. The ratio, M_w/M_n, previously called the polydispersity index (PDI) but recently renamed dispersity ($Ð$) by the International Union of Pure and Applied Chemistry (IUPAC),[6] indicates the statistical distribution of the molecular weight in a sample. Typically, a synthetic polymer sample has non-uniform dispersity, meaning that the sample has macromolecules of different molecular weights, while polymer samples, such as simple proteins, having macromolecules of the same molecular weights have uniform dispersity. Gel permeation or size-exclusion chromatography and

(a)

(b)

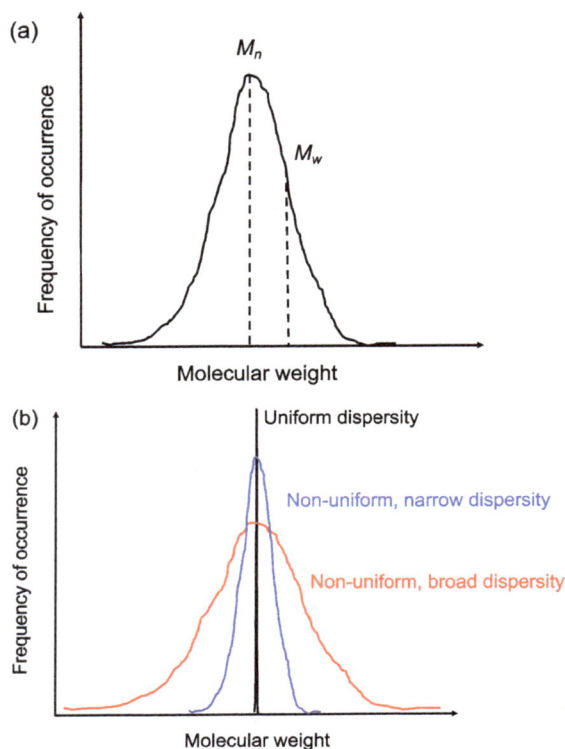

Figure 1.8 Hypothetical molecular weight distribution plot depicting (a) number- and weight-average molecular weights and (b) dispersity.

matrix-assisted laser desorption ionisation mass spectrometry can determine the Đ of polymers (discussed in Chapter 4). Another way to describe the size of a polymer is the degree of polymerisation, which is the average number of monomers incorporated into the polymer. As the degree of polymerisation increases, the molecular weight increases because more monomers are incorporated into the polymer chain.

1.6   Classification of Polymers

Polymers can be classified according to their structural and functional properties. While most of the classification criteria discussed in the following subsection apply to all polymers, a few, the function of metal atoms, are specific to inorganic and organometallic polymers.

1.6.1   Classification Based on Structural Properties

A polymer structure is an important characteristic not only because it impacts other properties but because it forms a criterion to classify polymers. For inorganic and organometallic polymers, structural features such as atom connectivity, dimensionality, tacticity, function of the metal within the polymer framework, and the homology and sequential arrangement of monomers within the polymer framework categorise the polymer.

1.6.1.1 Atom Connectivity

This refers to the number of atoms connected to a defined atom within the polymer backbone. An organometallic polymer containing a tungsten carbene side chain (Figure 1.9a),[7] for instance, has a connectivity of one because the metal is connected to one atom within the polymer backbone. The metallopolytriazolate (Figure 1.9b)[8] typifies a polymer with a connectivity of two because the platinum atom is conjugated to two different atoms within the polymer backbone structure. Connectivity in coordination and organometallic polymers can range from one for polymers with side chains to more than ten in polymers containing metal coordination or sandwich complexes (Figures 1.9 and 1.10). Mixed connectivity can be found in polymers bearing different inorganic elements in their repeat unit (Figure 1.11).

1.6.1.2 Dimensionality

Inorganic and organometallic polymers can also be described based on the minimum number of directions needed to identify an atom within the polymer backbone. A linear polymer (Figure 1.9a), for instance, is one-dimensional (1D) because an atom in the polymer can move only in one direction. A polymer nanosheet (Figure 1.10a) is two-dimensional (2D) since an atom in the polymer framework can move in two, x and y, directions. Three-dimensional (3D) polymers such as organometallic dendrimers (Figure 1.12) and 3D metal–organic frameworks are now widespread. Atoms in 3D polymers can be described using three coordinates, x, y, and z. By manipulating the dimensionality of polymers, their functional properties, such as the electron transport

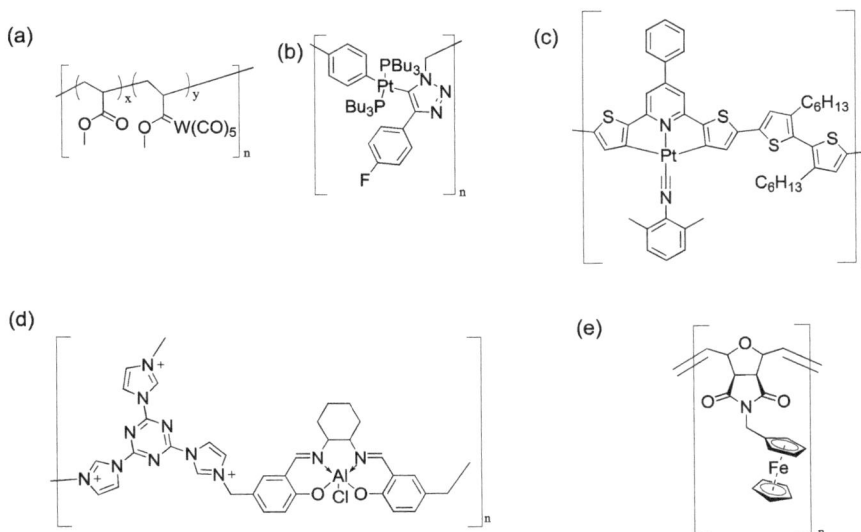

Figure 1.9 Schematics of coordination and organometallic polymers having a connectivity of (a) one, (b) two, (c) three, (d) four, and (e) five.

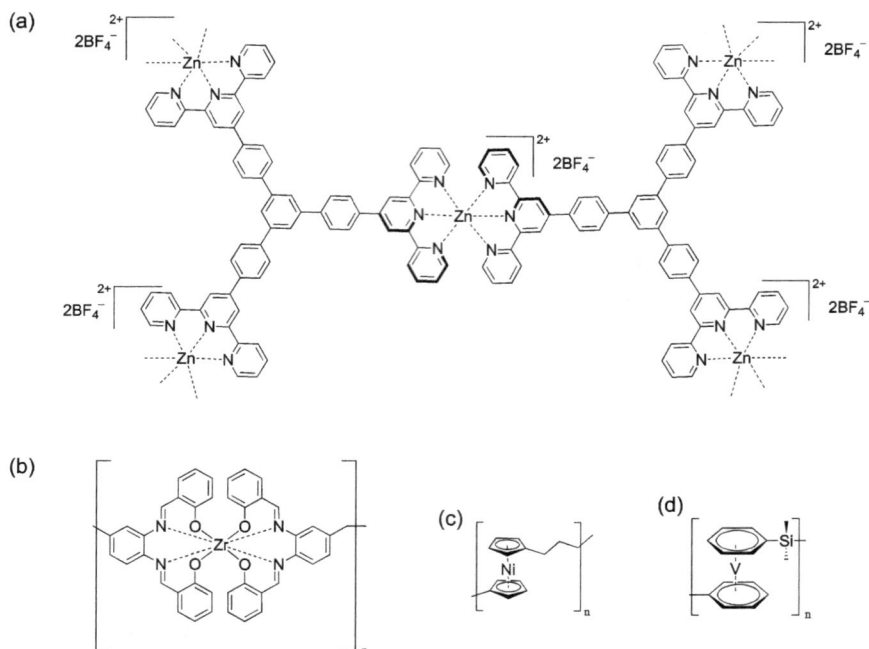

Figure 1.10   Schematics of coordination and organometallic polymers having a connectivity of (a) six, (b) eight, (c) ten, and (d) twelve.

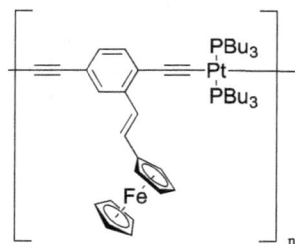

Figure 1.11   Schematics of organometallic polymers having a mixed connectivity of two and five.

properties of conducting polymers and the surface area of catalytic and luminescent polymers, can be fine-tuned (discussed in Chapter 5).

1.6.1.3 *Tacticity*

IUPAC defines tacticity as "*the orderliness of the succession of configurational repeat units in the main chain of a regular macromolecule*".[6] Different tacticities are possible in inorganic and organometallic polymers, and they define the relative configuration of substituent groups in adjacent chiral centres within the polymer. A structure where the substituent groups are on the same side of the chiral centres in the polymer backbone is termed isotactic (Figure 1.13a). If the substituents regularly alternate between the centres, it is termed syndiotactic, and if they are randomly oriented, it is called atactic (Figure 1.13b and d). Other configurations such as heterotactic (Figure 1.13c) exist, but

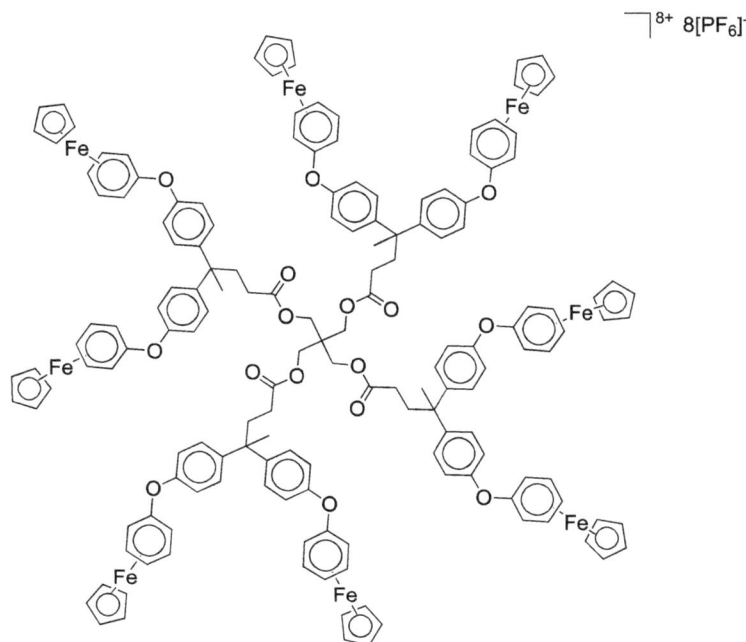

Figure 1.12 Schematic of an organometallic dendrimer, a 3D hyperbranched polymer.

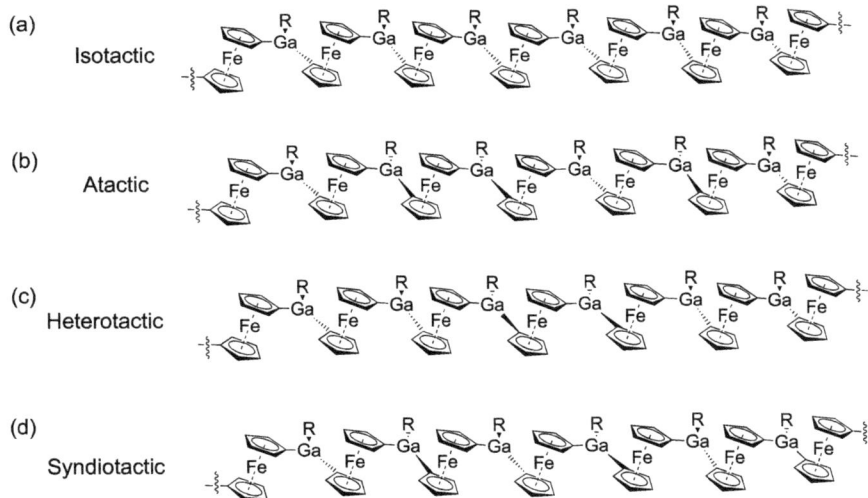

Figure 1.13 Possible tacticities in an organometallic polymer, gallium-bridged polyferrocene.

they have lesser commercial importance and are more challenging to synthesise than isotactic, syndiotactic, and atactic polymers. Heterotactic polymerisation requires a higher degree of stereocontrol because stereoselectivity needs to occur in an alternating manner. The importance of tacticity lies in its influence on crystallinity, which ultimately affects the mechanical properties, solubility, and glass transition and melting temperatures of polymers (discussed in Chapter 5). Isotactic and syndiotactic polymers are typically semi-crystalline, while atactic polymers are amorphous. The tacticity of polymers depends on the polymerisation conditions, such as temperature, solvent, catalyst, and monomer.

1.6.1.4 Function of the Metal Atom

Another criterion to classify organometallic polymers uses the function of the metal atom within the polymer framework. The classification results in two classes: metal-connected and metal-functionalised polymers. In the metal-connected polymer class, the metal atom is critical to the structure of the polymer backbone as exemplified by the nickel atom in the coordination polymer shown in Figure 1.14.[9] In the other class, metal-functionalised polymers, the metal is not essential to the structure of the polymer backbone but determines the functional properties such as redox and luminescent properties of the polymers. For example, the iron atom in the coordination polymer (Figure 1.14) is not critical to the structure of the polymer but determines its redox activity.[9] In metal-functionalised polymers, the metal atom can be conjugated or pendant to the polymer backbone but is immaterial to the structure of the backbone; it contributes to the overall macrostructure of the polymer.

1.6.1.5 Homology and Sequential Arrangement of Monomers

A homopolymer is derived from one type of monomer compared to a copolymer obtained by linking different monomers (Figure 1.15). In copolymers, the exact sequence of the monomers affects the structure and ultimately the properties of the polymer. At one extreme are block copolymers obtained by orderly linking a long block of one type of monomer, followed by another block of a different monomer, while at the other extreme are random copolymers obtained by arbitrarily conjugating the different monomers (Figure 1.16).

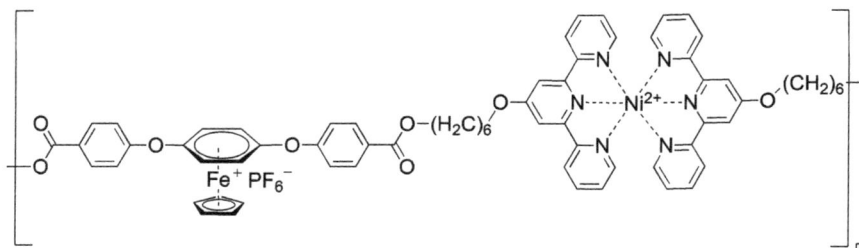

Figure 1.14 A schematic of an organometallic polymer having metal atoms that play structural and functional roles. The nickel atom is critical to the polymer backbone structure, while the iron atom contributes to the functional properties, specifically the redox activity of the polymer.

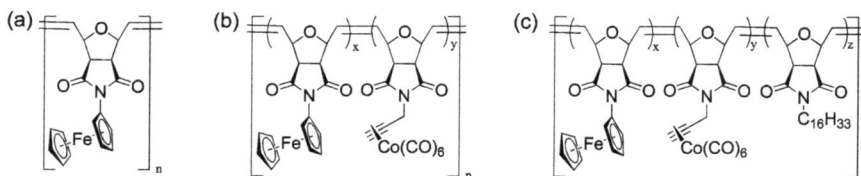

Figure 1.15 Schematics of a homopolymer (a) and copolymers (b and c).

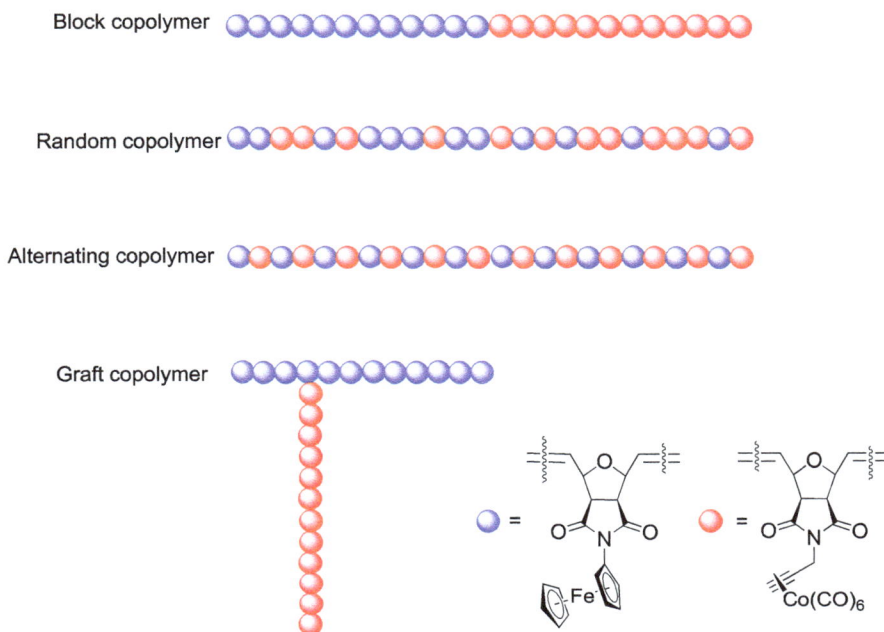

Figure 1.16 Schematic of the possible structures of copolymers.

An intermediate scenario exists where the co-monomers are linked in an alternating fashion to form an alternating copolymer (Figure 1.16). Graft copolymers are obtained by attaching one polymer to the backbone of another polymer (Figure 1.16). The monomer sequence in the polymer depends on polymerisation conditions such as the temperature, pressure, catalyst, and relative reactivity of the co-monomers. It should be noted that tacticity is a microstructural property of polymers, while homology and sequential arrangement of the monomers are macrostructural properties.

1.6.1.6 Polymer Architecture

Another macrostructural property of polymers that influences other properties is the architecture, which can be linear, branched, or network. A linear polymer consists of monomers linked in a linear sequence, while a branched polymer contains at least one branch point between the end groups of the main polymer backbone. Whereas all graft copolymers have branched architecture, not all branched polymers are graft copolymers because branched homopolymers exist. The unique chemistry of metals and organometallic compounds enables the design of many unusual, branched polymer architectures such as star polymers (Figure 1.17), comb polymers, ladder polymers, hyperbranched polymers, and dendrimers (Figures 1.12 and 1.17). Network polymers are derived by crosslinking linear or branched polymers through covalent or coordination bonds (Figure 1.17).

1.6.1.7 Elemental Composition

Polymers can be classified as inorganic, coordination, or organometallic depending on their elemental composition (Box 1.4). Inorganic polymers are derived wholly from

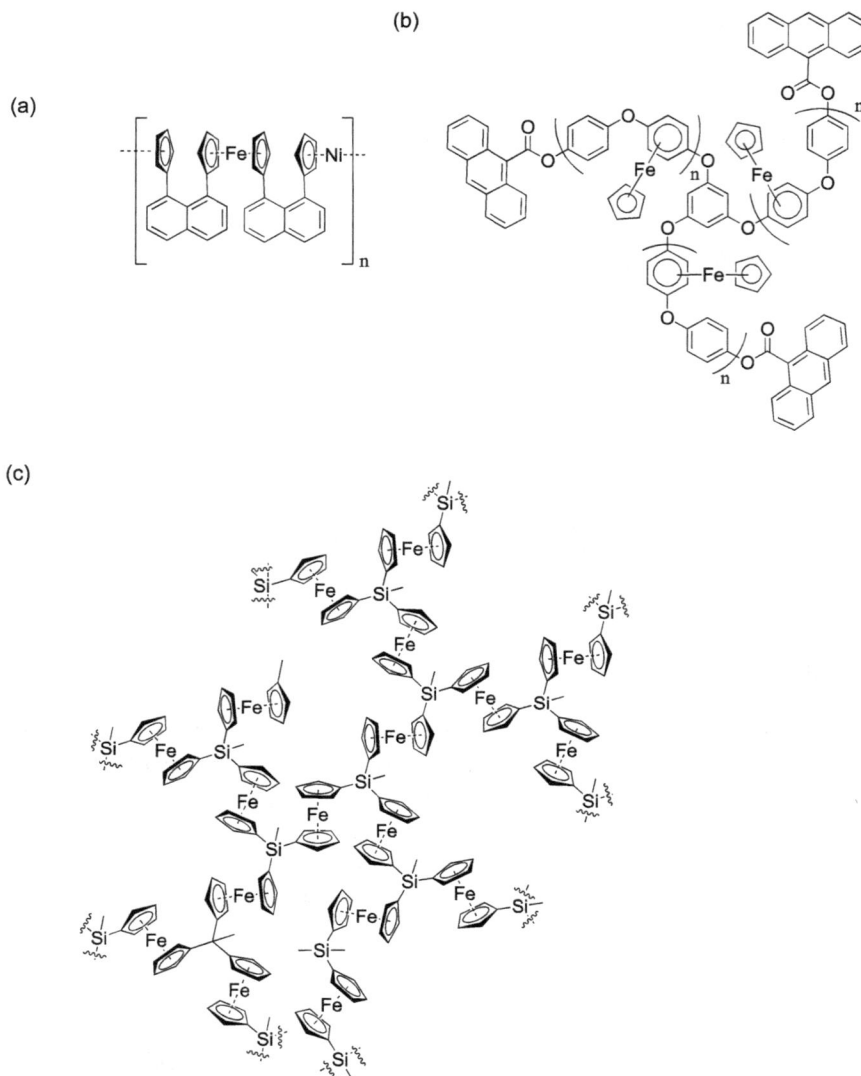

Figure 1.17 Schematics of different organometallic polymer architectures: (a) ladder, (b) star polymer, and (c) network polymer.

inorganic elements and contain no carbon atoms. Although some inorganic polymers exhibit excellent thermal stability, they generally feature low molecular weight, poor hydrolytic stability, and intractability to conventional polymer processing techniques. These properties make them undesirable precursors for the fabrication of films, fibres, and plastics. Typical examples include polysulphur, polymetaphosphate, and polyultraphosphate (discussed in Chapter 6). Organometallic and coordination polymers contain inorganic and organic elements. The organic component contributes to the rigidity of the structure and improves the solubility and processibility of the polymer, features that enable the synthesis of moderate to high molecular weight polymers. The difference between coordination and organometallic polymers is sometimes blurry. However, generally, coordination polymers are derived through coordination bonding between an organic ligand

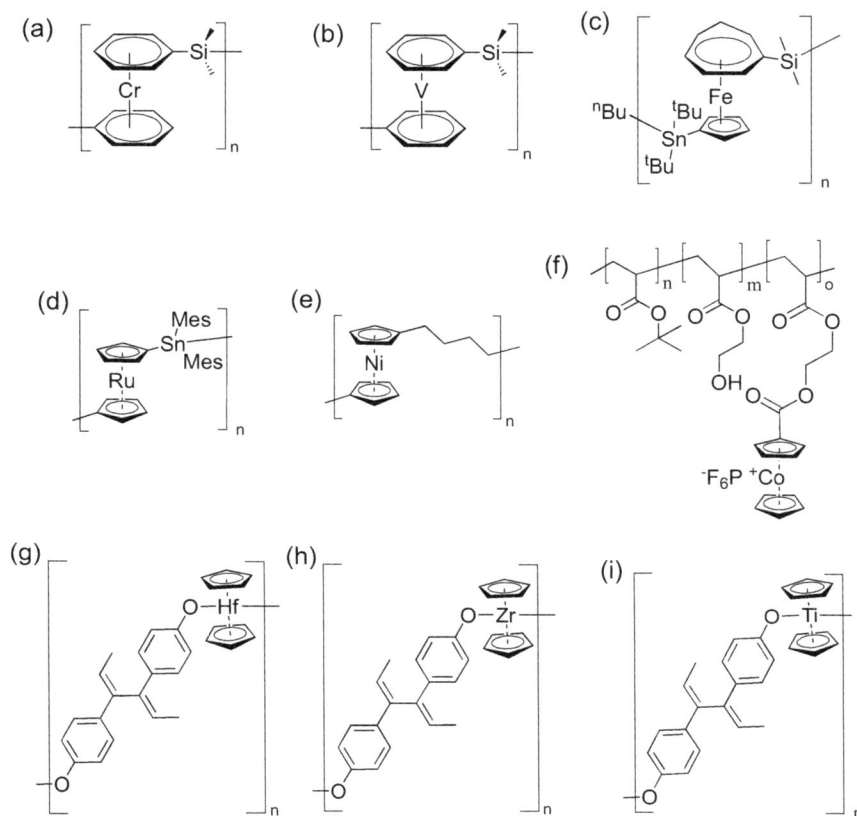

Figure 1.18   Examples of metallocene polymers.

and a metal atom, while in organometallic polymers the metal is linked to the organic moiety through a covalent bond. In coordination polymers, the denticity of the ligand and the spin state of the metal centre control the structure of the polymers, and eventually the degree of crosslinking, stability, and solubility. For instance, multidentate ligands such as the bis-tetradentate Schiff-base bridging ligand stabilise zirconium(IV) polymers and low-spin four-coordinate d^8 platinum(II) provides inertness to polymers containing non-aromatic bis-bidentate ligands.[10] Metallocene polymers are the pre-eminent example of organometallic polymers (discussed in Chapter 6). Today, various metallocene polymers derived from cobalt, titanium, ruthenium, and chromium have been synthesised (Figure 1.18), but like most metal-containing polymers they are plagued by poor stability and processibility and are yet to enter the market.[11]

1.6.2   Classification Based on Mechanical Properties

A processed polymer can be classified as an elastomer, a fibre, or a plastic depending on its mechanical behaviour, specifically its stress–strain relationship (Figure 1.19). This relationship is usually obtained by plotting the response of the polymer to deformation when force is applied to the point of permanent deformation. The stress–strain plot

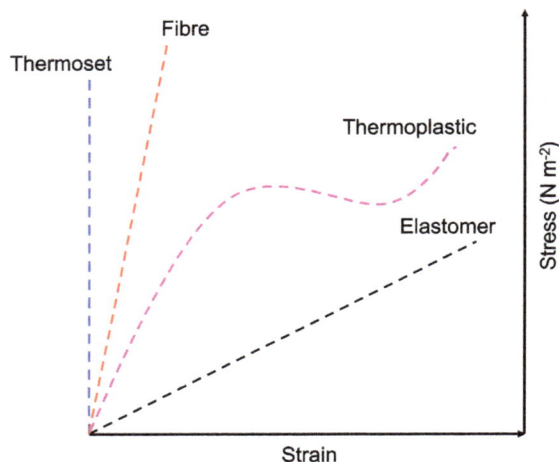

Figure 1.19 A hypothetical stress–strain curve of a thermoset, a fibre, a thermoplastic, and an elastomer.

Box 1.10 Mechanical properties of polymers.

1. Stress is the force applied to a material divided by the cross-sectional area.
2. Strain is deformation or displacement that results from stress.
3. Young's modulus is the resistance to deformation as determined by the slope of the stress–strain plot.
4. Yield strength defines the stress required to deform the polymer reversibly.
5. Tensile strength is the stress required to deform the polymer permanently.
6. Elastic elongation is the total elongation to the point of reversible deformation.
7. Ultimate elongation is the total elongation to the point of permanent deformation.

reveals important mechanical information, including Young's modulus, yield strength, ultimate or tensile strength, ultimate elongation, and elastic elongation, properties that qualify a polymer as an elastomer, a fibre, or a plastic (Box 1.10). Elastomers are polymers with low initial Young's modulus (<100 $N\,cm^{-2}$) and, under low stress, undergo extensive elastic elongation ($\leq 500\%$). They are typically amorphous polymers with a low glass transition temperature (the temperature at which a polymer undergoes a reversible transition from a brittle, glassy state to a viscous rubbery state) and weak intermolecular forces. Fibres are characterised by very high moduli ($>35\,000$ $N\,cm^{-2}$) and tensile strength ($>35\,000$ $N\,cm^{-2}$) and undergo low elongation ($<50\%$) under stress. They are highly crystalline polymers with high melting temperature (T_m) ($200\,°C \leq T_m \leq 300\,°C$) due to strong intermolecular forces, which results from the presence of polar functional groups within the polymer framework. Plastics are polymers that exhibit stress–strain behaviour that is intermediate between those of elastomers and fibres. Based on thermal processing behaviour, plastics are subclassified into thermoplastics, which can be heat-softened, and thermosets, which are resistant to heat-softening. Thermoplastics exhibit a moderate-to-high degree of crystallinity, moduli ($15\,000$–$350\,000$ $N\,cm^{-2}$), and ultimate elongation (20–800%), in contrast to thermosets which have extensive crosslinking and therefore are amorphous, highly rigid, and possess high moduli (70 000–350 000 N cm^{-2}) and small elongation (0.5–3%) under stress.

1.7 Naming Inorganic and Organometallic Polymers

It is acceptable to name polymers in honour of their history, as exemplified by Bakelite, named after its inventor, Leo Baekeland. Of course, IUPAC emphasises an unambiguous nomenclature that recognises the historical development of chemical compounds. However, with increasing scientific knowledge, IUPAC recommends "IUPAC names" for chemical compounds to promote unambiguous communication among scientists from different countries and reduces legal tussles over inventions. For organic polymers, two systems of naming are recommended: source-based and structure-based. In the source-based system, the polymer is named by attaching the prefix "poly" to the IUPAC name of its monomer. In the structure-based system, the name bears no relationship with the monomer but is based on the repeat unit structure of the polymer. The naming of regular single-strand inorganic and co-ordination polymers follows the same principles as their organic counterparts. For instance, in the structure-based system, the name of the polymer is based on the structure of the repeat unit prefixed by "poly" and other structural descriptors for end groups. The structural unit of the repeat unit and the end groups are named based on IUPAC rules for inorganic and coordination chemistry. Box 1.11 presents some general principles for naming inorganic and coordination polymers.[11] In addition to the source- and structure-based systems is the traditional naming system, commonly used in industry and academia. For instance, *catena*-poly[(dimethylsilicon)-μ-oxido] is commonly called polydimethylsiloxane, while

---

Box 1.11 IUPAC principles for naming inorganic and coordination polymers.

1. For polymers with unknown dimensionality, the name comprises the prefix "poly", followed by the IUPAC name of the structural unit of the repeat unit enclosed in square brackets. For example, poly[repeat unit].
2. For a linear, one dimensional polymer, the italicised prefix "*catena*" is added to the name of the polymer. For example, *catena*-poly[repeat unit].
3. When possible, specify the end groups of the polymer by adding prefixes "α" and "ω" to the name of the polymer. For example, α-(end group)-ω-(end group)-*catena*-poly[repeat unit].
4. Regular, single-strand polymers are named by following the principles listed above. For example, if the repeat unit is a homoatomic structural unit like sulphur, then name the polymer by citing the mononuclear central atom prefixed by "poly". For example, *catena*-poly[sulphur] (Table 1.2).
5. If there are side chains on the mononuclear central atom, cite the atom together with the side chain as a ligand. For example, *catena*-poly[diethyltin] (Table 1.2).
6. If the repeat unit consists of a mononuclear central atom and one bridging ligand, name the polymer as described in rules 4 and 5, but followed by the name of the bridging ligand prefixed by "μ". For example, *catena*-poly[(ammine-chloridozinc)-μ-chlorido] and *catena*-poly[titanium-tri-μ-chlorido] (Table 1.2).

7. If there is more than one element in the repeat unit, the most senior element is the central atom, and the other element is considered the bridging ligand. For example, *catena*-poly[(diphenylsilicon)-μ-oxido] and α-ammine-ω-(amminedichloridozinc)-*catena*-poly[(amminechlorido-zinc)-μ-chlorido] (Table 1.2).

8. For metallocene-containing polymers (Table 1.2), use rules for organic nomenclature. The seniority rules for selecting the central atom for organic polymers differ from those in inorganic elements.

9. For guidance on how to determine the senior element, please consult the IUPAC rules in *Pure Appl. Chem.*, 1985, **57**, 149–168.

Table 1.2 IUPAC names of some inorganic and organometallic polymers.

Polymer	IUPAC name
S⸢S⸣$_n$S	*catena*-Poly[sulphur]
[-Sn(CH$_3$)(CH$_3$)-]$_n$	*catena*-Poly[dimethyltin]
[Ti(Cl)(Cl)(Cl)]$_n$	*catena*-Poly[titanium-tri-μ-chlorido]
[Ag-NC]$_n$	*catena*-Poly[silver-μ-(cyanido-N:C)]
[-Si(C$_6$H$_5$)(C$_6$H$_5$)-O-]$_n$	*catena*-Poly[(diphenylsilicon)-μ-oxido]
H$_3$N-[Zn(NH$_3$)(Cl)-Cl-Zn(NH$_3$)-Cl]$_n$	α-Ammine-ω-(amminedichloridozinc)-*catena*-poly[(amminechloridozinc)-μ-chlorido]
[-Si(CH$_3$)(CH$_3$)-ferrocene-Fe-]$_n$	Poly[(dimethylsilanediyl)ferrocene-1,1-diyl]

poly[(dimethyl-silanediyl)ferrocene-1,1-diyl] is commonly called *catena*-polyferrocenylsilane (Box 1.11). In this monograph, we will use the traditional names to describe inorganic and organometallic polymers (Boxes 1.12 and 1.13).

Box 1.12   Worked Example 1.4.

Name the following polymers by the IUPAC recommended system:

  a.

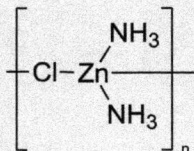

  b.

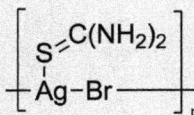

  c.

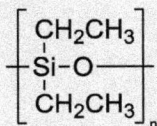

  d.

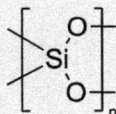

Solution:

  Refer to the IUPAC recommended rules in Box 1.5

  a. *catena*-poly[(diamminezinc)-μ-chlorido]
  b. *catena*-poly[[(thiourea-S)silver]-μ-bromido]
  c. *catena*-poly[(diethylsilicon)-μ-oxido]
  d. *catena*-poly[silicon-di-μ-oxido]

Box 1.13   Review questions.

  1. A 10 g sample of hypothetical polymers A–E shown below was fractionated into three samples of different molecular weights given in the following table.

    a. Estimate the number-average and weight-average molecular weights (hints: recall eqn (1.1) and (1.3) and the relationship between mass and mole).
    b. Compute the dispersity.
    c. Classify the polymerisations into condensation, addition, step, and chain.
    d. Classify the polymers into copolymers, homopolymers, and inorganic, organometallic, and coordination polymers.

e. Which polymer is likely to exhibit redox and optical activity?

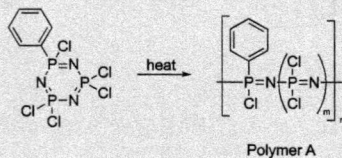

Polymer A

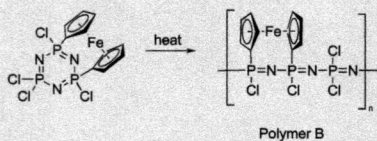

Polymer B

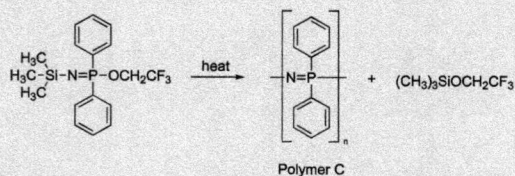

Polymer C

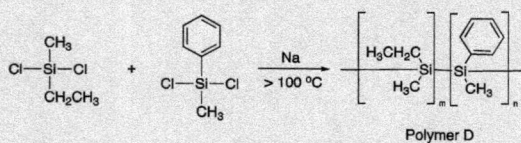

Polymer D

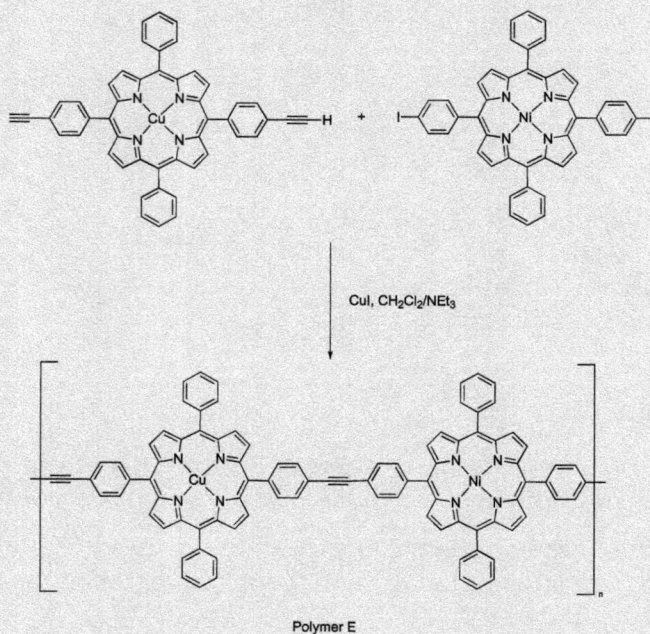

Polymer E

Polymer	Mass (grams)			Molecular weight ($g\,mol^{-1}$)		
	Fraction 1	Fraction 2	Fraction 3	Fraction 1	Fraction 2	Fraction 3
Polymer A	4	2	4	58 000	76 000	45 000
Polymer B	6	3	1	98 000	66 000	34 000
Polymer C	5	2	3	23 000	12 000	44 000
Polymer D	2	6	2	95 000	150 000	135 000
Polymer E	4	3	3	77 000	45 000	110 000

Further Reading

1. D. Feldman, Polymer History, *Des. Monomers Polym.*, 2008, **11**, 1–15.
2. F. Furukawa, *Inventing Polymer Science: Staudinger, Carothers, and the Emergence of Macromolecular Chemistry*, University of Pennsylvania Press, Philadelphia, 1998.
3. S. R. Batten, S. M. Neville and D. R. Turner, *Coordination Polymers: Design, Analyses, and Application*, Royal Society of Chemistry, Cambridge.
4. A. S. Abd-El-Aziz, C. Agatemor and N. Etkin, *Macromol. Rapid Commun.*, 2014, **35**, 513.
5. J. E. Sheats, History of Organometallic Polymers, *J. Macromol. Sci. A*, 1981, **15**, 1173–1199.
6. J. E. Mark, H. R. Allcock and R. West, *Inorganic Polymers*, Oxford University Press Inc., New York, 2005.
7. R. D. Archer, *Inorganic and Organometallic Polymers*, Wiley-VCH, New York, 2001.
8. G. Odian, *Principles of Polymerization*, John Wiley & Son Inc., Hoboken, 2004.
9. International Union of Pure and Applied Chemistry, Nomenclature for Regular Single-Strand and Quasi Single-Strand Inorganic and Coordination Polymers, *Pure Appl. Chem.*, 1985, **57**, 149–168.

References

1. F. Arimoto and A. Haven Jr, *J. Am. Chem. Soc.*, 1955, **77**, 6295.
2. H. Rosenberg and E. W. Neuse, *J. Organomet. Chem.*, 1966, **6**, 76.
3. D. A. Foucher, B. Z. Tang and I. Manners, *J. Am. Chem. Soc.*, 1992, **114**, 6246.
4. W. H. Carothers, *J. Am. Chem. Soc.*, 1929, **51**, 2548.
5. P. J. Flory, *Principles of Polymer Chemistry*, Cornell University Press, 1953.
6. R. G. Jones, J. Kahovec, R. Stepto, E. S. Wilks, M. Hess, T. Kitayama and W. V. Metanomski, Polymer Division, in *Compendium of Polymer Terminology and Nomenclature: IUPAC Recommendations, 2008*, International Union of Pure and Applied Chemistry, Royal Society of Chemistry, Cambridge, 2009, vol. 464.
7. B. J. Wilson and J. N. Brantley, *J. Am. Chem. Soc.*, 2019, **141**, 12453.
8. C. Beto, E. Holt, Y. Yang, I. Ghiviriga, K. Schanze and A. Veige, *Chem. Commun.*, 2017, **53**, 9934.
9. A. S. Abd-El-Aziz, J. L. Pilfold, B. Z. Momeni, A. J. Proud and J. K. Pearson, *Polym. Chem.*, 2014, **5**, 3453.
10. R. D. Archer, *Inorganic and Organometallic Polymers*, John Wiley & Sons, 2004, vol. 4.
11. N. G. Connelly, T. Damhus, R. Hartshorn and A. Hutton, *Nomenclature of Inorganic Chemistry: IUPAC Recommendations 2005*, Cambridge, Royal Society of Chemistry, 2005.

2 Step Polymerisation

2.1 Introduction

This chapter focuses on the synthesis of inorganic and organometallic polymers *via* step polymerisation. Generally, the synthesis of an inorganic or organometallic polymer involves the polymerisation of monomers. However, in some instances, post-synthesis metalation – conjugation with metals – of organic polymers yields organometallic or co-ordination polymers. We must mention that, compared to organic polymers, the synthesis of inorganic and organometallic polymers is more intellectually challenging. A fundamental difficulty is the lack of a unifying principle to guide the synthesis of polymerisable inorganic or organometallic monomers. This dilemma contrasts with that in organic synthesis, where the reaction outcomes are relatively more predictable based on concepts such as electronegativity, steric hindrance, inductive effects, and resonance effects. Also, unlike organic synthesis, where a few essential skillsets are enough to master the art, a wide range of techniques, including Schlenk techniques, is required to practise inorganic and organometallic synthesis. This more challenging situation, which is also evident in chain polymerisation (discussed in Chapter 3), is a consequence of the large number of atoms and structures encountered in inorganic chemistry. This chapter introduces synthesis strategies used in the step polymerisation of inorganic and organometallic monomers. This chapter also provides insights into how to relate some fundamental chemistry principles such as solvent polarity, nucleophilicity, and electrophilicity with the step polymerisability of inorganic and organometallic monomers. We will present principles to guide the student in designing inorganic and organometallic polymers (Box 2.1).

2.2 Fundamentals of Step Polymerisation

Condensation reactions such as esterification, amidation, and etherification can be modified to synthesise inorganic and organometallic polymers. These reactions

Fundamentals of Inorganic and Organometallic Polymer Science
By Christian Agatemor, Kajal Ghosal, Samuel Fura and Peter J. S. Foot
© Christian Agatemor, Kajal Ghosal, Samuel Fura and Peter J. S. Foot 2024
Published by the Royal Society of Chemistry, www.rsc.org

Box 2.1 Learning outcomes.

By the end of this chapter, the student should be able to:

1. Explain step and chain polymerisations.
2. Describe the thermodynamic and kinetic factors that affect step polymerisation.
3. Predict the effects of inorganic and organometallic groups on the efficiency and mechanism of step polymerisation.
4. Propose mechanisms for the step polymerisation of inorganic and organo-metallic monomers.
5. Design main-chain or side-chain inorganic and organometallic polymers using step polymerisation.

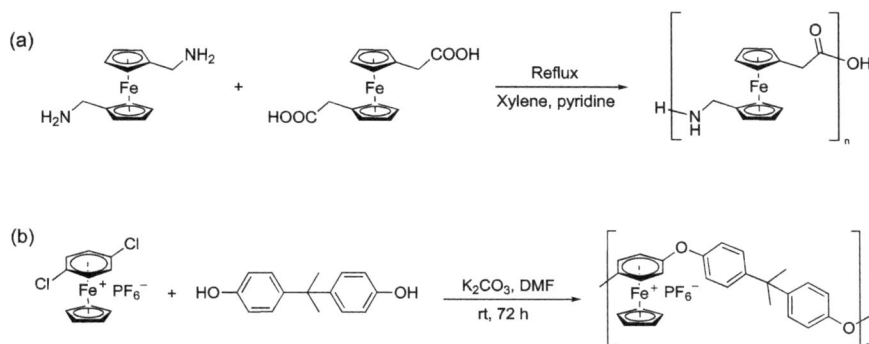

Figure 2.1 Step polymerisation involving two bifunctional monomers.

proceed *via* a step mechanism and generally yield a small-molecule condensation byproduct alongside the target polymer, and hence they are often also called poly-condensations or step polymerisations. Generally, step polymerisation occurs be-tween polyfunctional monomers to yield a polymer with or without a small-molecule byproduct. Step polymerisation can be subclassified into two groups. In one group, two different polyfunctional monomers that bear different functional groups react to form a polymer. Typical examples of this group of step polymerisation include the amidation reaction between 1,1′-di(aminomethyl)ferrocene and 1,1′-di(carboxy-methyl)ferrocene to form an iron-containing polyamide (Figure 2.1a). Another example is the etherification reaction between bisphenol A and η^6-dichlorobenzene-η^5-cyclopentadienyliron(II) hexafluorophosphate to give an iron-containing polyether (Figure 2.1b). In the other group, the polymerisation involves a single monomer that possesses the same or different functional groups. A classic example of this group is the synthesis of an iron-containing polyamide from the amidation reaction of 1-carboxymethyl-1′-aminomethylferrocene (Figure 2.2a) or a tin-containing polyphthalocyanine from the etherification reaction of dihydroxyl tetra-*t*-butylphthalocyanato (Figure 2.2b). Step polymerisation yields practically useful, high molecular weight polymers only at very high monomer conversions (>98–99%), which is achievable under stringent conditions (Box 2.2).

(a)

(b)

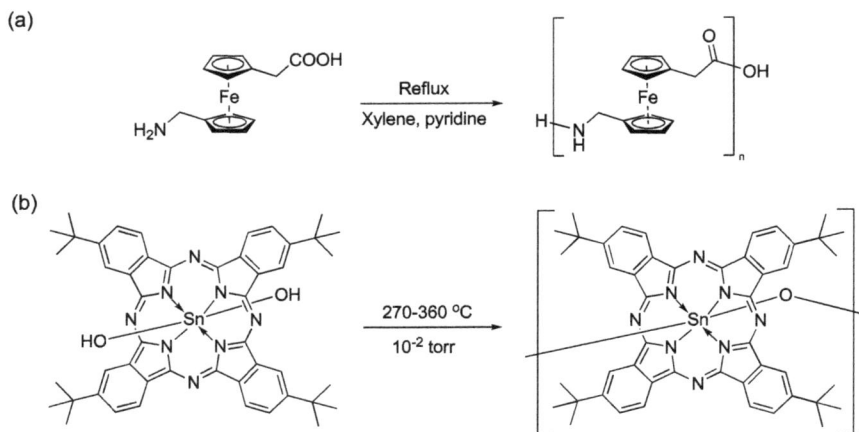

Figure 2.2 Step polymerisation involving a single bifunctional monomer.

Box 2.2 Strategy to increase the molecular weight of step polymers.

Step polymerisation typically produces low molecular weight polymers; however, high molecular weight polymers can be obtained if the following conditions are imposed on the reaction system:

1. Eliminate the small molecule byproduct from the reaction vessel.
2. Reduce side reactions such as cyclisation to favour polymer formation.
3. Control the reaction equilibrium to drive the polymerisation towards polymer formation.
4. Use a precise stoichiometric amount of the reacting functional groups.

2.2.1 Kinetic Aspects of Step Polymerisation

Step polymerisation proceeds with a relatively slow increase in polymer molecular weight. Nearly 99% of the monomers are consumed before the formation of practically useful, high molecular weight polymers (discussed in Chapter 1). At any time, the reaction mixture consists of monomers and small-, intermediate-, and large-sized polymeric species. The reactivity of the functional groups on these polymeric species is independent of molecular weight, meaning the rate of step polymerisation is virtually the same as that of an analogous condensation reaction between small molecules that bear the same functional groups. Thus kinetics does not explain the widely observed decreasing polymerisation rate with increasing polymer molecular weight. The observed decrease in rate is partly due to reduced polymer solubility as the molecular weight increases and not the decreased reactivity of the functional groups. Theoretically, high molecular weight polymers exhibit lower diffusion rates, but the reactivity of the functional group depends on the collision frequency, which remains unchanged due to conformation rearrangement of the segments of the growing polymer chain.[1] Conditions where the frequency of collision of the functional group decreases can occur when the molecular weight becomes exceedingly high, leading to a diffusion-controlled polymerisation rate. Under these conditions, polymer diffusion becomes slow and inadequate to maintain a collision frequency to sustain a reaction.[1] The practical implication of this kinetics is that polymerisation conditions such as solvent, temperature, and concentration that impact polymer solubility also affect the polymerisation rate. On the other hand, the same conditions may introduce side reactions such as cyclisation and degradation, which compete with the synthesis of high molecular weight polymers. Therefore, a chemist must optimise the polymerisation conditions for achieving an optimum polymerisation rate and minimal deleterious side reactions (Box 2.3).

Box 2.3   Worked example 2.1.

Question

A hypothetical step polymerisation of 1,1'-di(hydroxymethyl)ferrocene and 1,1'-di(carboxymethyl)cobaltocenium hexafluorophosphate yields only low molecular weight polymers.

   a. Depict the polymerisation with a chemical equation.
   b. Predict the byproduct.
   c. Provide strategies to increase the molecular weight of the polymer.

Answer

   a.

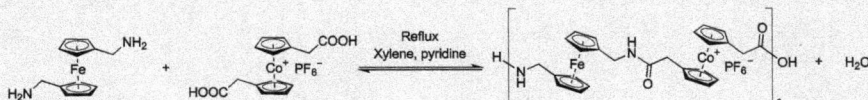

   b. The byproduct is water.
   c. The polymerisation is an equilibrium reaction, with the forward reaction being polymerisation and the reverse one being a depolymerisation reaction. Factors that favour the forward reaction should increase the molecular weight of the polymer. These factors include removing the water byproduct by purging the reaction vessel with inert gas, reducing the reaction pressure, or increasing the reaction temperature close to or above the boiling point of water. It is crucial to ensure that the high temperature does not degrade the polymer. Preferably, a solid desiccant or molecular sieve can be used to absorb the water. Also, water can be trapped and deactivated by using a solvent such as toluene, which forms a toluene/water azeotrope, or dimethyl sulfoxide, which has a high affinity for water. Finally, new monomers can continuously be added to the reaction vessel.

2.2.2 Thermodynamic Perspectives of Step Polymerisation

Step polymerisations are equilibrium reactions involving forward and reverse reactions. The forward reaction is polymerisation, while the reverse reaction is a depolymerisation reaction. An important question to consider is: will an equilibrium step polymerisation yield high molecular weight polymers if nothing is done to drive the reaction towards the polymer side? Step polymerisations produce a small-molecule byproduct such as water or hydrogen chloride in addition to the growing polymer chain. The small molecule provides a favourable positive entropy that drives the polymerisation forward in opposition to the negative entropy that results from the growing polymer chain. In a closed system – one in which there is no transfer of reactants into or products out of the reaction vessel – the reaction eventually reaches equilibrium, where the polymerisation rate equals the depolymerisation rate. The extent to which the forward reaction proceeded before reaching equilibrium determines the molecular weight of the polymer. Therefore, the logical strategy to obtain high molecular weight polymers is to carry out step polymerisation in an open system, meaning new monomers are continually added or one of the products is continually removed from the reaction vessel. In most step polymerisations, the small molecule rather than the polymer is removed to drive the forward reaction towards the synthesis of high molecular weight polymers. A volatile small molecule byproduct can be removed using high temperature, reduced pressure, or a purge with inert gas. For acidic byproducts, such as hydrogen chloride, the most popular method involves adding a base, either as a homogeneous solution or as a heterogeneous solid, to the reaction vessel to neutralise the acid. A combination of these methods can be applied to drive the equilibrium towards high molecular weight polymers. In the step polymerisation of diphenolic nucleophiles with η^6-dichloro-benzene-η^5-cyclopentadienyliron(II) hexafluorophosphate to give iron-containing poly(aromatic ethers) *via* nucleophilic aromatic substitution (Figure 2.1b), potassium bicarbonate, a base, is added to neutralise the eliminated hydrogen chloride byproduct. Alternatively, the reaction can be carried out at 60 °C to enable the diffusion of hydrogen chloride out of the vessel.

2.2.3 Effect of Metal and Ligand Chemistry

An important factor to consider during the synthesis of inorganic polymers, specifically coordination polymers, is the metal and ligand chemistry. The electronic configuration and coordination geometries of the metal atom and the chelating effect of the coordinating ligand are pivotal to the stability and utility of a polymer. Inert metals, such as the platinum group metals, yield the most stable and practically useful polymers. However, unless the polymerisation is conducted under special conditions, these metals are not reactive as monomers and have a weak affinity towards the bridging ligand. In contrast, labile metals are reactive as monomers, but the resulting polymers can quickly degrade or dissociate into short oligomers unless a suitable solvent or coordinating ligand is used to stabilise the polymer. In some instances, the solvent promotes the degradation of the polymer. For instance, polysilanes are synthesised by the reductive coupling of diorganodichlorosilanes with alkali metals in nonpolar solvents such as toluene and xylene heptane, or octane at high temperature, usually above 100 °C (Figure 2.3).[2]

M is an alkali metal

Figure 2.3 Modified Wurtz coupling of dichlorosilane to polysilane.

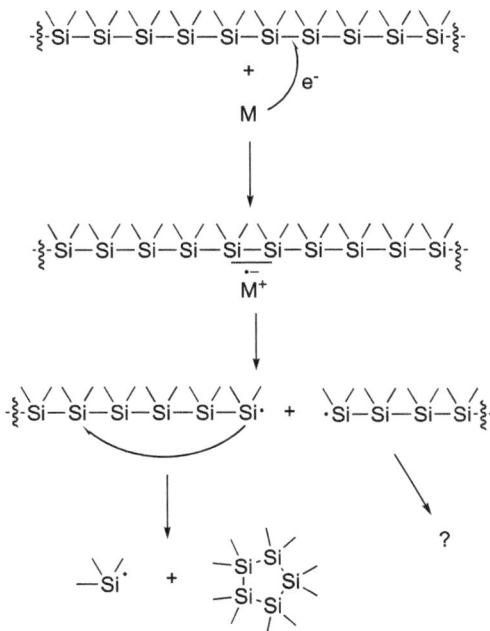

Figure 2.4 Degradation of polysilane mediated by alkali metal reaction with the backbone silicon atoms.

Under these conditions, no substantial degradation occurs, but in polar solvents such as tetrahydrofuran or diglyme, which solvate alkali metal ions, polysilanes undergo degradation to cyclic oligomers.[2] The degradation rate depends on the temperature, type of alkali metal, solvent polarity, and substituents on the silicon atom. In tetrahydrofuran, a moderately polar solvent, poly(methylphenylsilylene), degrades in less than 10 minutes to the corresponding cyclopentasilanes, whereas no degradation occurs in toluene, a nonpolar solvent. The mechanism of the degradation involves the formation of a polymeric radical anion, a process that is facilitated by the energy gained from the solvation of alkali metal cations.[2] The radical anion dissociates into a radical and a silyl anion that reacts with inter- and intramolecular Si–Si bonds. The intramolecular reaction is faster due to entropic effects and leads to the formation of cyclic oligomers (Figure 2.4).

The coordination geometry of the metal contributes to polymer solubility. Reduced solubility leads to premature precipitation, which removes the growing polymer from the reaction mixture, leading to oligomers that may lack practical utility. Coordination polymers containing square planar metal ions such as Ni^{2+} coordinating bis(bidentate) bridging ligands, such as oxalate ions, tend to stack like aromatic polymers, resulting in premature precipitation.[3] The presence of multiple electron pairs on the donor atoms of the ligand also contributes to polymer intractability by forming interchain metal–ligand interactions.[3] Several synthesis strategies can be used to enhance the solubility of a growing polymer to enable the synthesis of high molecular weight polymers (Box 2.4).

2.3 Types of Step Polymerisation

In the following subsections, we will consider the various condensation reactions adapted for the synthesis of inorganic and organometallic polymers.

2.3.1 Nucleophilic Aromatic Substitution

Aryl halides cannot undergo classic nucleophilic substitution (S_N2 and S_N1) reactions but become susceptible to nucleophilic aromatic substitution (S_NAr) reactions when coordinated to transition metal complexes. This susceptibility results from the electron-withdrawing properties of the metal, which activates the arene nucleus towards nucleophilic attack and makes the halide substituent a good leaving group. For example, coordinating transition metal complexes such as $Cr(CO)_3$, $Fe(C_5H_5)$, $Ru(C_5H_5)$, or $Mn(CO)_3$ to a dichloroarene ligand (Figure 2.5) renders the chloro group a good leaving group, enabling the synthesis of many metal-containing polymers under mild reaction conditions. The coordination of the cyclopentadienyliron ligand $Fe^+(C_5H_5)$ to the dichlorobenzene ligand yields η^6-dichlorobenzene-η^5-cyclopentadienyliron(II)

Box 2.4 Strategies to enhance the solubility of inorganic and organometallic polymers.

1. Use a bulky ligand to perturb the stacking geometry and preclude interchain metal–ligand interactions.
2. Change the geometry of the complex, for example, from a square planar to an octahedral or a tetrahedral metal centre.
3. Use a suitable polar solvent to solvate the metal centre.

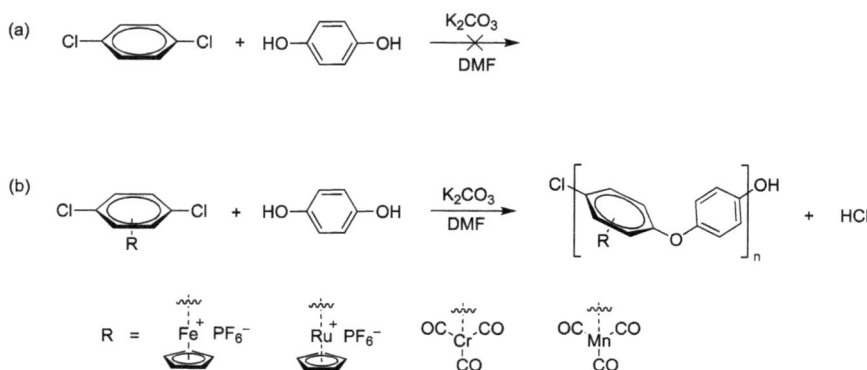

Figure 2.5 (a) Aryl chlorides do not undergo nucleophilic substitution reactions (b) but do when complexed with organometallic compounds.

hexafluorophosphate, which reacts by S_NAr with diphenols, dithiols, or diamines to give iron-containing poly(aromatic ethers), poly(thioethers) or poly(amines), respectively (Figures 2.6–2.8).[4] S_NAr-mediated step polymerisation is only successful with alcohols if the nucleophile is a tertiary carbanion, limiting this method to the syntheses of poly-(aromatic ethers). In contrast, primary, secondary, and tertiary thiols successfully undergo S_NAr-mediated step polymerisation to yield polythiols. With phenols and thiols, potassium bicarbonate is usually added to the reaction vessel to neutralise the hydrogen

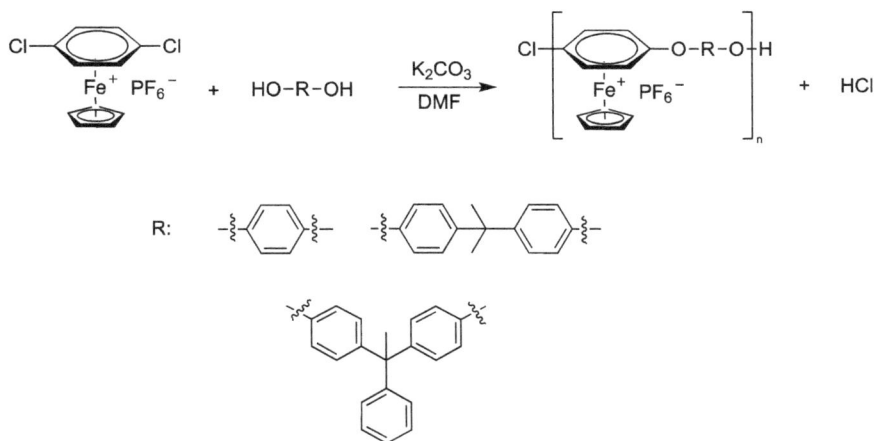

Figure 2.6 Step polymerisation *via* a nucleophilic aromatic substitution reaction between η^6-dichlorobenzene-η^5-cyclopentadienyliron(II) hexafluorophosphate and a diphenol yields an iron-containing poly(aromatic ether).

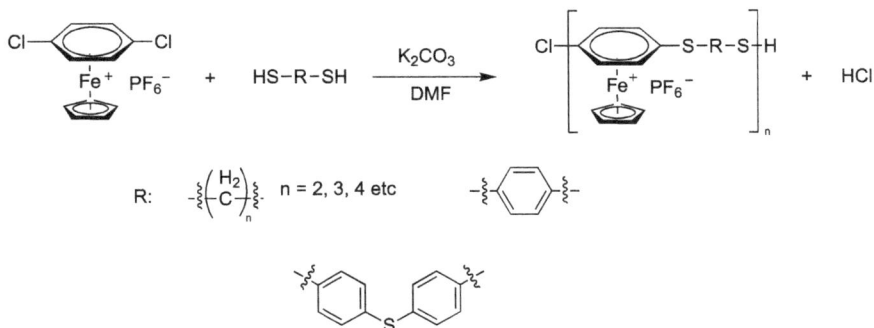

Figure 2.7 Step polymerisation *via* a nucleophilic aromatic substitution reaction between η^6-dichlorobenzene-η^5-cyclopentadienyliron(II) hexafluorophosphate and a dithiol yields an iron-containing poly(thiol ether).

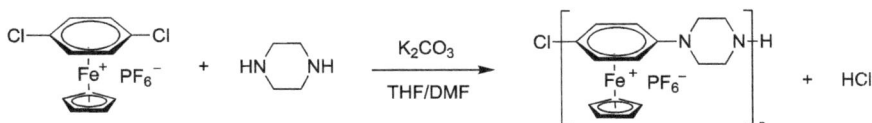

Figure 2.8 Step polymerisation *via* a nucleophilic aromatic substitution reaction between η^6-dichlorobenzene-η^5-cyclopentadienyliron(II) hexafluorophosphate and a di-amine yields an iron-containing polyamine.

chloride byproduct eliminated during the polymerisation. While polymerisation is attainable at room temperature, it is common to run the reaction at 60 °C to increase the polymerisation rate and eliminate hydrogen chloride to enable the synthesis of high molecular weight polymers.

S_NAr-mediated step polymerisation affords polymers of different architectures. Dinucleophiles such as bisphenol A react with η^6-dichloroarene-η^5-cyclopentadienyliron(II) to yield linear polymers, while trinucleophiles (Figure 2.6) such as benzene-1,3,5-triol give star and hyperbranched polymers (Figure 2.9). The method also allows the synthesis of

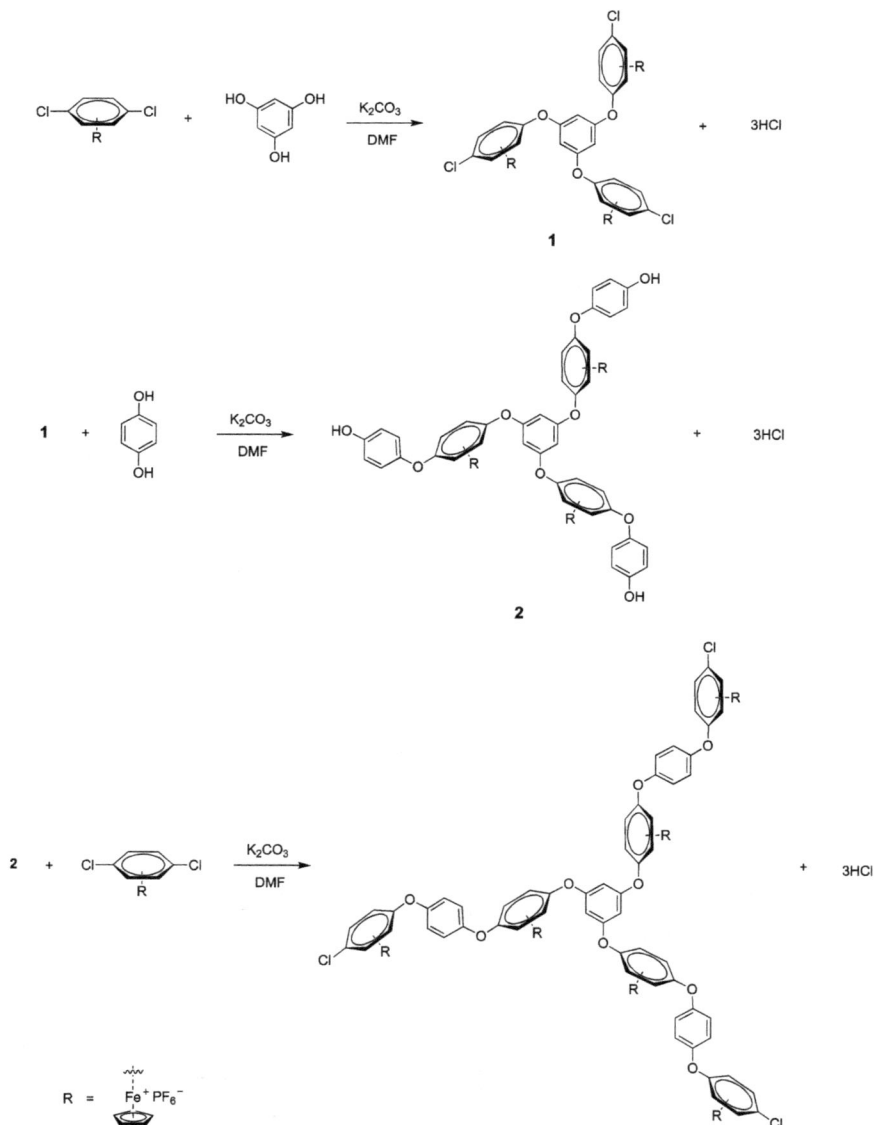

Figure 2.9   Synthesis of an iron-containing star polymer based on a nucleophilic aromatic substitution reaction of a trinucleophile, such as benzene-1,3,5-triol, and η^6-dichlorobenzene-η^5-cyclopentadienyliron(II) hexafluorophosphate.

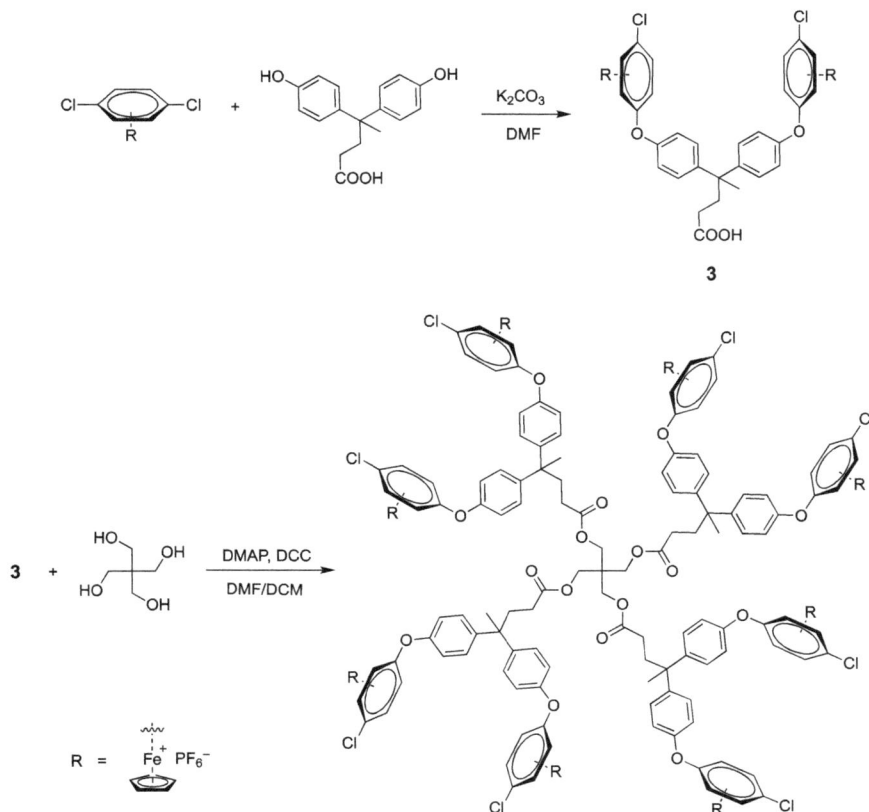

Figure 2.10 Synthesis of an iron-containing dendrimer *via* nucleophilic aromatic substitution reaction involving η^6-dichlorobenzene-η^5-cyclopentadienyliron(II) hexafluorophosphate and a phenol.

dendrimers (Figure 2.10) bearing η^6-dichloroarene-η^5-cyclopentadienyliron(II) moieties in their branches or peripheries, and copolymers of thiols, ethers, and amines (Figures 2.11 and 2.12).[4] Again, this polymerisation method provides a synthesis method for homometallic polymers, containing only iron complexes, and heterometallic polymers, containing different transition metals.[5]

Alkylating the arene ligand allows control of fundamental properties such as polymer solubility and thermal stability. Polymers derived from alkylated arenes feature superior solubility compared with their non-alkylated analogs. The transition metals enhance solubility in polar solvents such as dimethyl sulfoxide and dimethylformamide that can solvate the metal. However, in most cases, the method yields intractable polymers, which are insoluble in many standard laboratory solvents, including dichloromethane and tetrahydrofuran (Box 2.5).

2.3.2 Coupling Reactions

Transition metal-catalysed coupling reactions are a direct approach to create a bond between two molecules. These reactions, which can be classified into traditional coupling, reductive coupling, and oxidative coupling, are increasingly used to

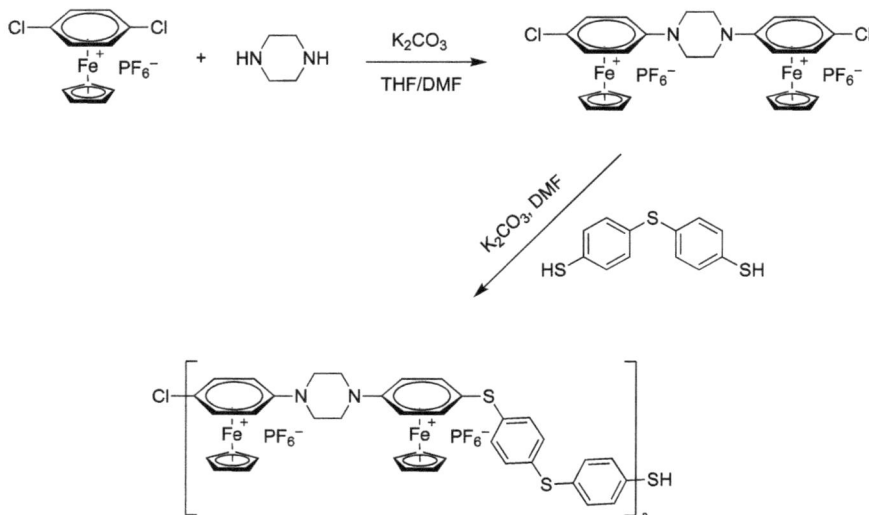

Figure 2.11 Synthesis of an iron-containing polyamine-*co*-polythiol *via* a nucleophilic aromatic substitution reaction involving diamine, dithiol, and η^6-dichlorobenzene-η^5-cyclopentadienyliron(II) hexafluorophosphate.

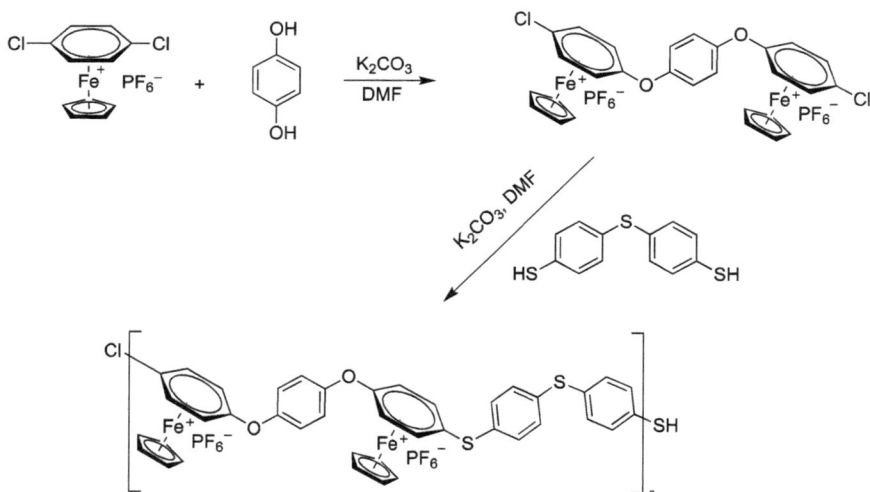

Figure 2.12 Synthesis of an iron-containing polyether-*co*-polythiol *via* S_NAr reaction-mediated step polymerisation involving diphenol, dithiol and η^6-dichlorobenzene-η^5-cyclopentadienyliron(II) hexafluorophosphate.

synthesise organometallic and inorganic polymers *via* a step polymerisation mechanism. Many coupling reactions have been explored for polymerisation, but they can broadly be classified into homocoupling which couples two identical molecules (Figure 2.13a) and heterocoupling, also known as cross-coupling, which combines two different molecules (Figure 2.13b). Both couplings can be used to synthesise inorganic and organometallic polymers, but cross-coupling is more widely explored. We will discuss some of the coupling reactions used for the polymerisation of inorganic and organometallic monomers.

Box 2.5 Worked example 2.2.

Question

Assuming all hypothetical reagents are commercially available, design and synthesise a heterobimetallic block copolymer using S_NAr reaction-mediated step polymerisation. Use chemical equations to illustrate your answer.

Answer

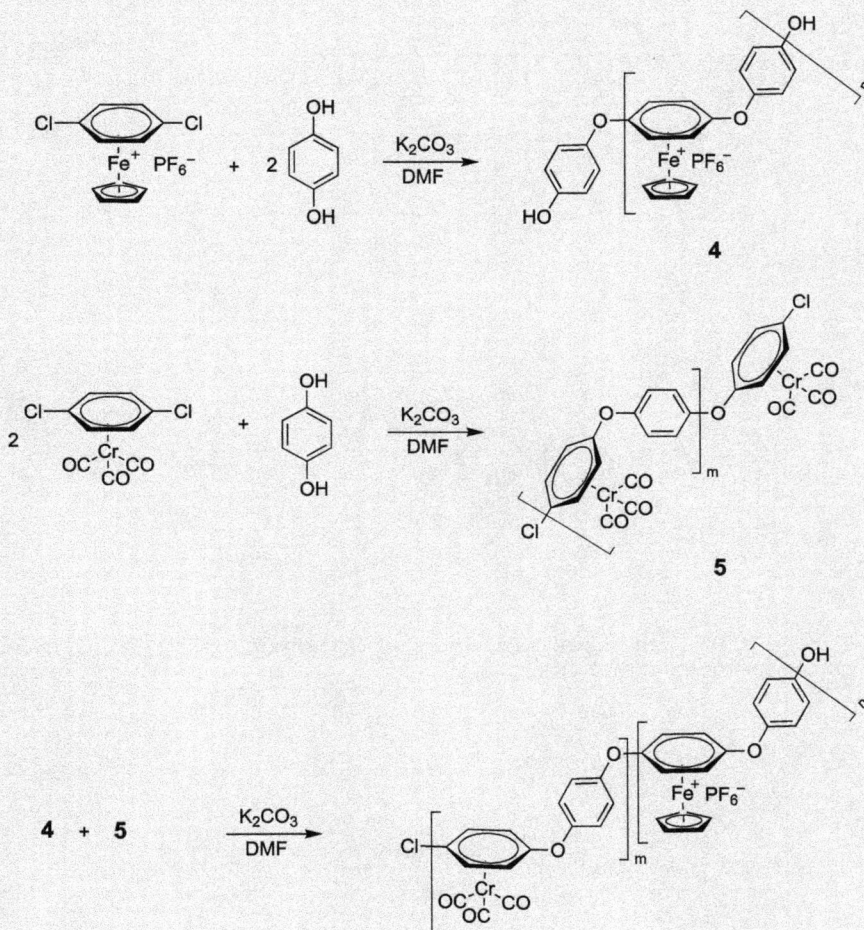

2.3.2.1 *Sonogashira Cross-coupling Polymerisation*

Sonogashira cross-coupling reaction is commonly employed to synthesise diverse organometallic polymers, including polymetallaynes, an attractive class of transition metal-containing conjugated polymers investigated as semiconductors for optoelectronic devices (discussed in Chapter 6). Polyplatinyne, a polymetallayne that incorporates platinum in the main chain of the polymer, is synthesised *via* Sonogashira cross-coupling reaction of alkyne-terminated monomers with platinum chlorides

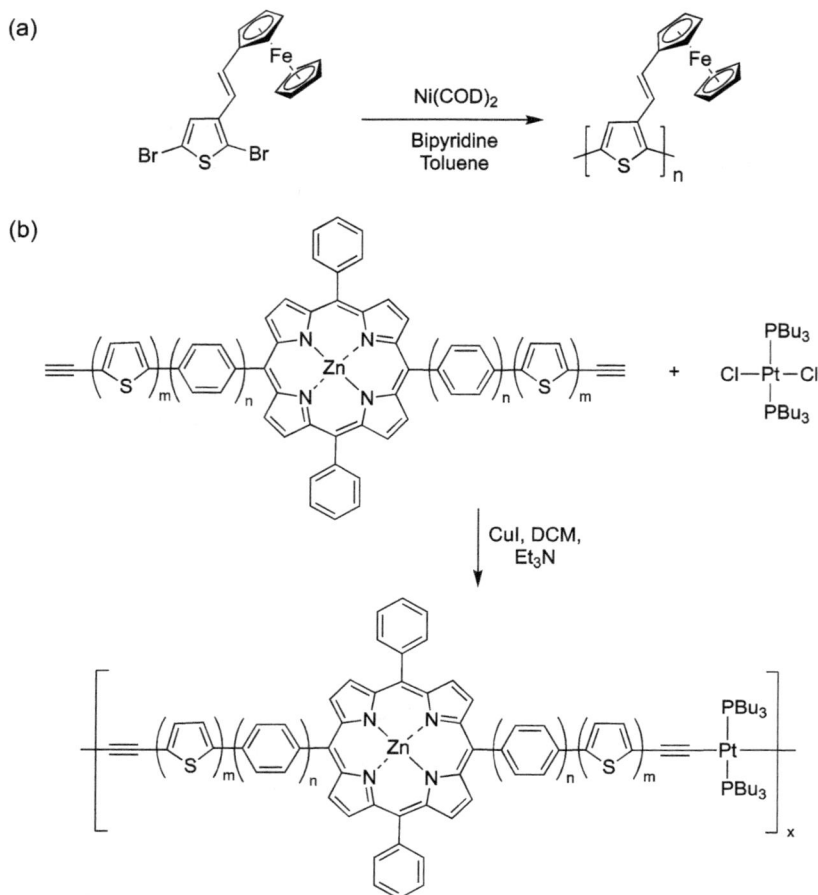

Figure 2.13 Synthesis of organometallic polymers *via* (a) homocoupling and (b) cross-coupling of organometallic monomers.

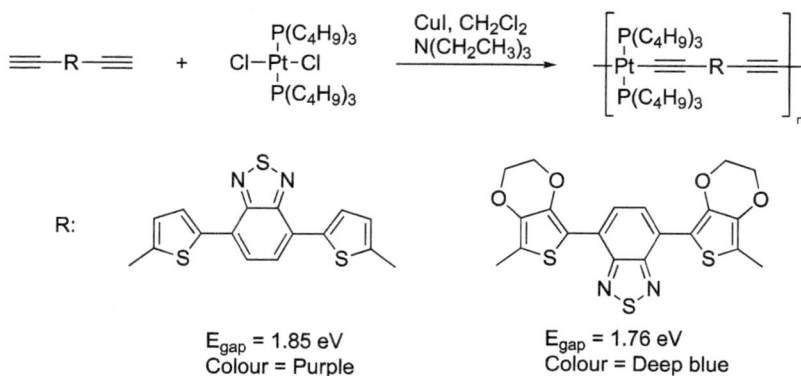

Figure 2.14 Properties of polyplatinynes can be tuned by changing the spacer between platinum moieties.

(Figure 2.13b). Like most step polymerisations, Sonogashira-type polymerisation results in a small-molecule byproduct, a hydrogen halide, which must be eliminated from the reaction system to drive the equilibrium towards high molecular weight polymers. The elimination of the byproduct is usually achieved by adding a base to the reaction to neutralise the hydrogen halide. Amines, such as diethylamine and triethylamine, are primarily used as solvents to deprotonate the hydrogen halide, but potassium carbonate or caesium carbonate is occasionally used to achieve the same results. The strength of the base affects the molecular weight of the polymer. For example, the cross-coupling reaction of 1,4-dihexyl-2,5-diiodobenzene with 1,3-diethynylcyclobutadiene-(cyclopentadienyl)cobalt in diisopropylamine ($pK_a = 10.63$) yields only low molecular weight polymers, but the use of a stronger base, piperidine ($pK_a = 11.22$), instead resulted in higher molecular weight polymers (Figure 2.14).[6]

Sonogashira cross-coupling polymerisation affords copolymers and heterometallic polymers (Figure 2.13b) and is quite versatile concerning the number of transition metals incorporated into the polymer framework. For example, the reaction provides routes to incorporate transition metals such as mercury, iridium, cobalt, titanium, ruthenium, and gold into the main chain of polymers.[7] Sonogashira cross-coupling tolerates various structural modifications on the monomers. Therefore, polymers with various structural modifications can be designed to target a specific fundamental or functional property (discussed in Chapter 6). Changing the structure of the aromatic spacer in a polyplatinyne from 3,4-ethylenedioxythiophene-benzothiadiazole[8] to 4,7-di-2′-thienyl-2,1,3-benzothiadiazole[9] alters the optoelectronic properties (discussed in Chapter 6), precisely the color and bandgap of the polymer from deep blue to purple and 1.76 eV to 1.85 eV, respectively (Figure 2.14). Sonogashira cross-coupling polymerisations are feasible at high (up to 145 °C), room, and cold (down to −75 °C) temperatures in a deaerated or ambient medium.

2.3.2.2 Acyclic Diyne Metathesis Polymerisation

Acyclic diyne metathesis (ADIMET) polymerisation is an alternative to palladium-catalysed cross-coupling and offers superior polymerisation yields, higher molecular weights and well-defined polymers. For example, the molecular weight of cobalt-containing poly-(*p*-phenylene-ethynylene) obtained *via* ADIMET polymerisation (Figure 2.15) is superior to

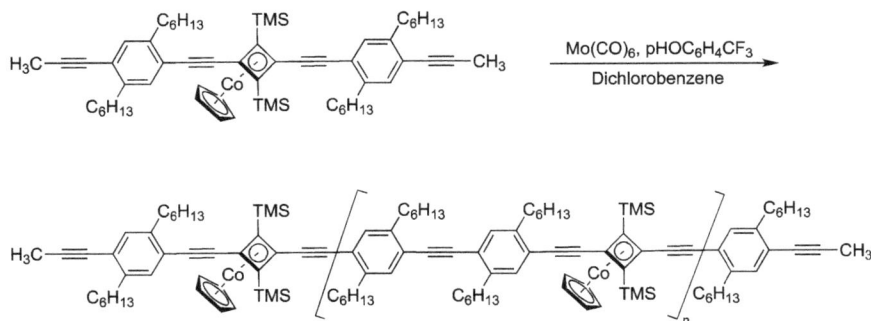

Figure 2.15 Acyclic diyne metathesis polymerisation yields a cobalt-containing poly(*p*-phenylene-ethynylene).

that obtained *via* palladium-catalysed Sonogashira cross-coupling.[10] In ADIMET polymerisation, the catalyst is formed *in situ*, and chain propagation to high molecular weight polymers is driven by eliminating 2-butyne instead of hydrogen halide. To achieve a high degree of polymerisation for the cobalt-containing poly(*p*-phenylene-ethynylene) (Figure 2.15), it is critical to use a high concentration of the molybdenum hexacarbonyl catalyst. A downside of ADIMET polymerisation is that it requires high temperatures (~ 150 °C) in contrast to the mild temperature condition of Sonogashira cross-coupling polymerisation.

2.3.2.3   Suzuki Coupling Polymerisation

Suzuki cross-coupling reaction involves organoboranes and organohalides. When adopted for polymerisation, the reaction provides a synthesis route to transition metal-containing conjugated polymers, such as platinum- and iridium-containing polymers. For instance, Suzuki cross-coupling reaction of 9,9-dioctyl-2,7-dibromofluorene, 9,9-dioctylfluorene-2,7-bis(trimethyleneborate), and (5-bromo-2-(4-bromophenyl)pyridine))-Ir(III)bis(2-phenylpyridine) gives an iridium-containing polymer (Figure 2.16).[11]

Suzuki cross-coupling polymerisation features many advantages over other cross-coupling polymerisations, such as the Stille cross-coupling reaction which uses potentially toxic organotin monomers. A prominent advantage is that the organoboranes used in the Suzuki cross-coupling reaction are safer and less toxic for the environment, inert to water and oxygen, and can tolerate various functional groups. These features are attractive because they make the polymerisation relatively eco-friendly and more economical. Unlike Sonogashira cross-coupling (discussed in Section 2.3.2.1), where the added base only neutralises the hydrogen halide byproduct, the base in a Suzuki cross-coupling polymerisation has three functions. The primary function of the base in Suzuki cross-coupling is to increase the transmetalation rate by converting the organoboronic acid to an organoborate, which is more reactive toward the organopalladium complex (Figure 2.17). The base also consumes boric acid, a reaction byproduct that perturbs the acid–base equilibrium, altering the reaction rate and selectivity. Therefore, the base indirectly ensures that the reaction proceeds toward high molecular weight polymers. In some instances, the molecular weight depends on the catalyst used in the cross-coupling

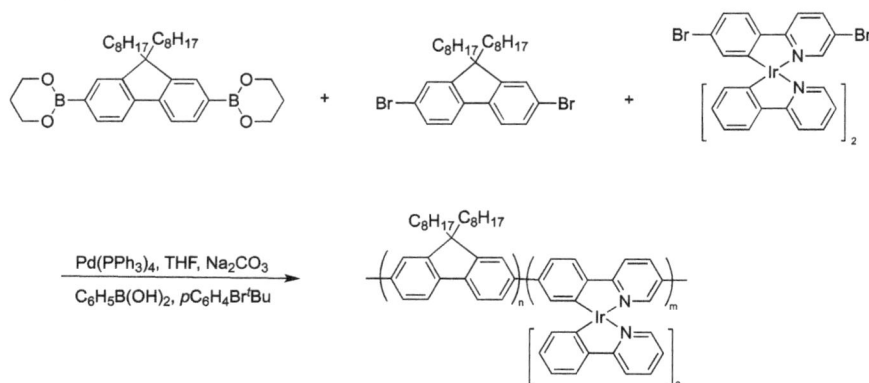

Figure 2.16   Suzuki coupling reaction yields an iridium-containing copolymer.

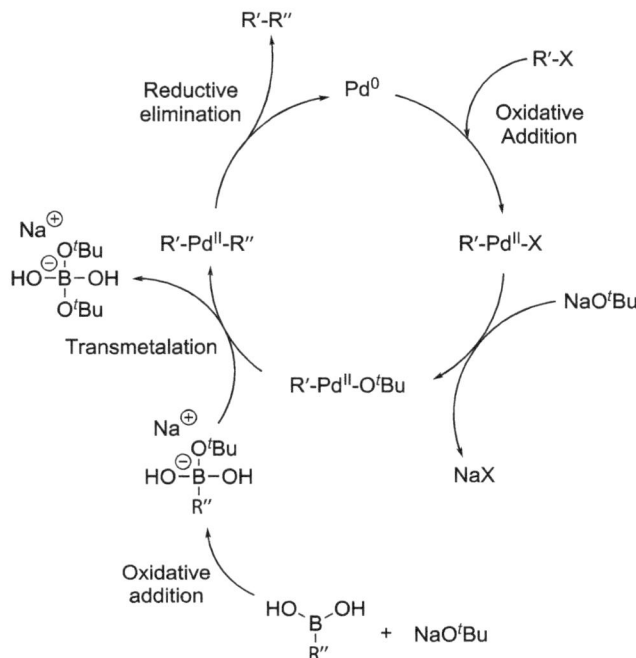

Figure 2.17 Mechanism of Suzuki cross-coupling showing the role of the base in the reaction.

polymerisation. For instance, a platinum-containing monomer that degrades when the polymerisation was carried out with the conventional Pd(PPh$_3$)$_4$ catalyst was less prone to degradation with the bulkier and more nucleophilic (Pd(P^{t}Bu$_3$)$_2$).[12] The bulkier catalyst is less likely to perturb the structural integrity of the monomer, allowing polymerisation to proceed because bulkiness will likely hinder coordination of the catalyst ligand to the platinum centre of the monomer. Significantly, for the catalyst, nucleophilicity increases the rate of the oxidative addition step, while spatially bulky structures promote orbital overlap with the metal to increase the rate of the reductive elimination step. Increasing the rates of both oxidative addition and reductive elimination steps promotes polymerisation to high molecular weight polymers at a much lower temperature (40 °C), compared with the typical 80–100 °C used in conventional palladium catalysts. In addition to Suzuki coupling polymerisation, the palladium catalyst bearing a bulky ligand (Pd(P^{t}Bu$_3$)$_2$) catalyses Stille coupling polymerisation for the synthesis of high molecular weight platinum-containing polymers.[12] The lower temperature also mitigates the degradation of the platinum polymer during polymerisation.

2.3.3 Bridging Ligand Coordination Polymerisation

Coordination polymers can be designed by taking advantage of the ability of transition metals to complete their coordination sphere through binding to various ligands. The ligand can be a monodentate ligand such as alkoxide, carboxylate, sulphide and acetylide ions or a multidentate ligand like terpyridines and Schiff bases. Metal–ligand coordination is achievable through many reactions, including nucleophilic substitution or salt elimination reactions. A nucleophilic substitution reaction between a transition metal phthalocyanine dichloride and a diphenol or a salt elimination reaction between a transition metal phthalocyanine dichloride and the sodium salt of *p*-diethynylbenzene results in a transition metal-containing phthalocyanine polymer bearing a phenoxide or acetylide ion coordinated metal centre, respectively (Figure 2.18).[13] A nucleophilic substitution or salt elimination reaction facilitates the polymerisation of several dihaloorganometallic molecules. These molecules react with acetylides, thiols, dicarboxylates, diamines, or diols to yield low molecular weight intractable polymers alongside small molecule byproducts

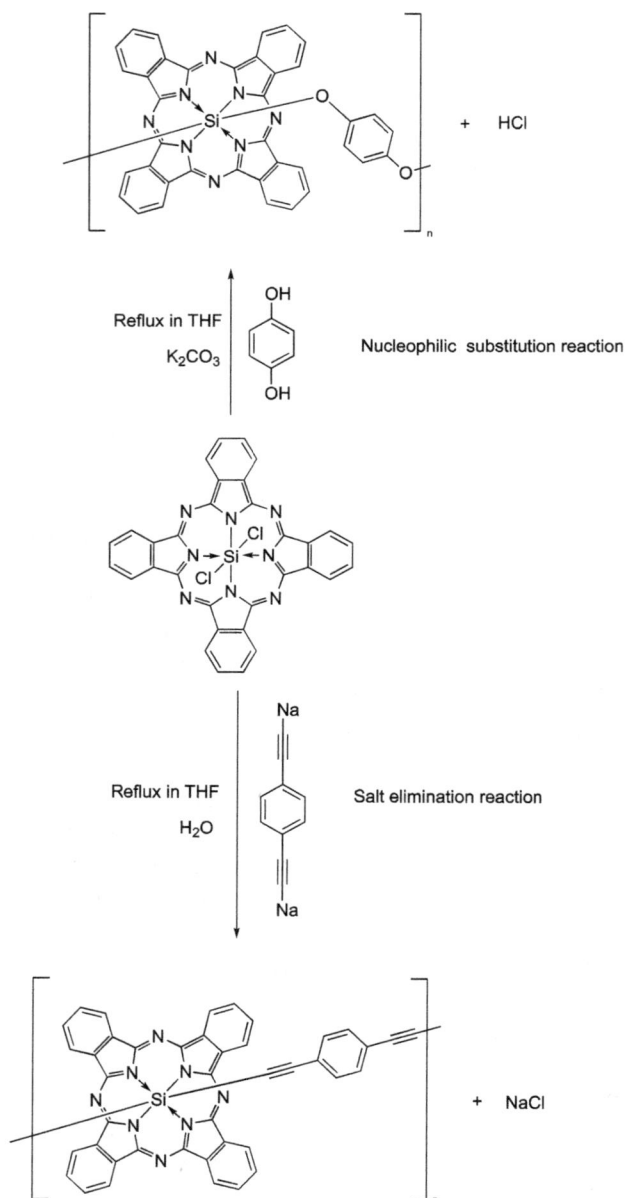

Figure 2.18 Synthesis of coordination organometallic polymers *via* nucleophilic substitution and salt elimination reactions.

such as hydrogen chloride or sodium chloride. Low molecular weight polymers are mainly produced due to premature precipitation of the growing polymers, byproducts perturbing the reaction equilibrium, and polymer decomposition due to metal reactivity. High molecular weight polymers are still attainable through a rational selection of the monomers. For example, the polymerisation depicted in Figure 2.19 involves the reaction of dienestrol and metallocene derived from metals with a d^0 electronic configuration (titanium(IV), zirconium(IV), and hafnium(IV)).[14] The successful synthesis of these high molecular weight metallocene polymers is due to the large charge and small size of these metal ions which provide inertness to support the stability of the polymers. The molecular weight of the polymers decreases in the order: titanocene > zirconocene > hafnocene (Table 2.1), apparently due to decreasing charge density, which decreases the inertness of the transition metals in their low (+2) oxidation state. Titanocene and zirconocene polymers are stable in hexamethylphosphoramide for three months, whereas the hafnocene polymer is stable for only two months under similar conditions.[14]

Multidentate ligands are better than monodentate ones in stabilising coordination polymers. The terpyridine ligand, reputed for its ability to coordinate various transition metal ions, typifies a multidentate ligand used to synthesise several coordination polymers. Terpyridine is a p-acceptor ligand that stabilises first-row transition metals, such as iron, cobalt, nickel, and zinc, in their low oxidation state, enabling the synthesis of terpyridine coordination polymers of various dimensionalities, ranging from 1D linear polymers through 2D nanosheets to 3D metal–organic frameworks. As another kind of multidentate ligand, tetradentate salen ligands, also known as Schiff bases, are noted for their ability to

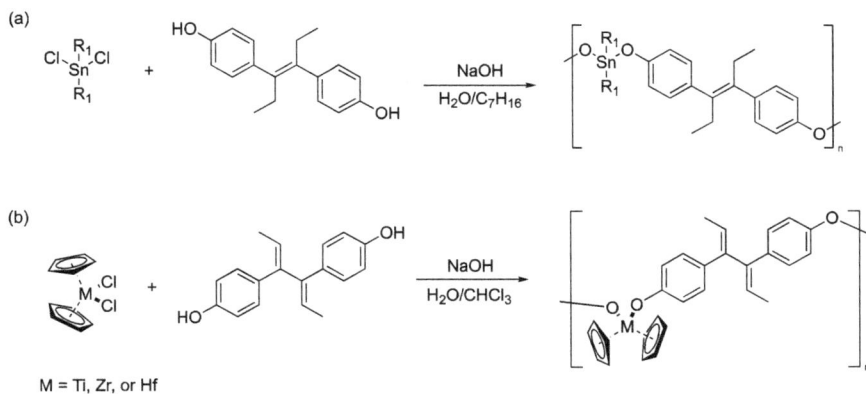

Figure 2.19 Nucleophilic substitution reaction yields organotin (a) and metallocene-containing polymers (b).

Table 2.1 The molecular weight and stability of metallocene-containing polymers.

Metallocene	M_w (Da)	Degree of polymerisation	Stability in HMPA[a]
Titanocene	3.3×10^7	70 000	3 months
Zirconocene	6.7×10^6	14 000	3 months
Hafnocene	2.1×10^5	400	2 months

[a] Hexamethylphosphoramide.

coordinate and stabilise various metals, including labile d^n and f^n ions in various oxidation states. The polymerisation depicted in Figure 2.20 involves a solvothermal reaction between a salen ligand and cerium alkoxide or acetylacetonate to form coordination polymers alongside the alcohol or β-diketone byproduct.[3] In these polymerisations, the tetradentate salen bridging ligands stabilise the cerium ion and the resulting polymer. Also, the hot dimethyl sulphoxide solubilises the organometallic monomer, ligand, and growing polymer chain to enable the synthesis of high molecular weight polymers.

Solvothermal-based condensation reactions are a classic method to synthesise metal–salen coordination polymers. Several transition metal polyelectrolytes are synthesised *via* the solvothermal process as exemplified by the copper(ɪ)/copper(ɪɪ)-salen coordination polymer, which is stable in air, water, ethanol, acetonitrile, and dimethylformamide (Figure 2.21).[15] The inertness of the aliphatic salen ligand provides an opportunity to integrate the solvothermal condensation reaction with other synthetic methodologies to design advanced metal–aliphatic salen polymers for functional applications such as heterogeneous catalysis. Also, post-polymerisation metallation of a salen polymer affords

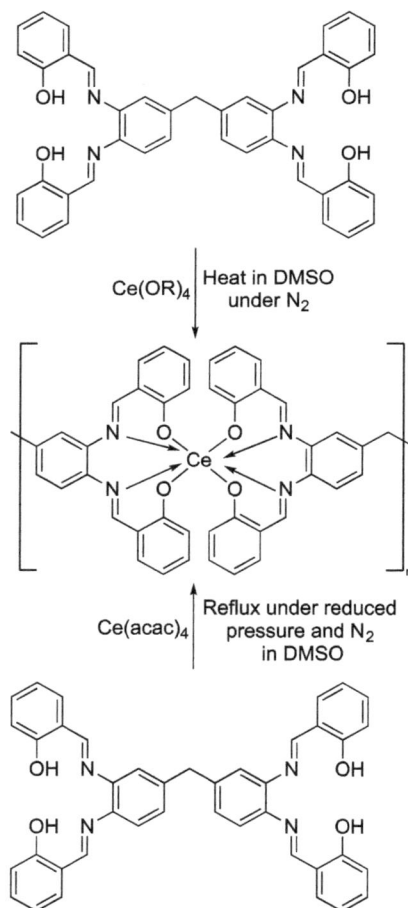

Figure 2.20   Synthesis of the cerium-containing polymer *via* the bridging ligand coordination method. The ligand used in this synthesis is salen.

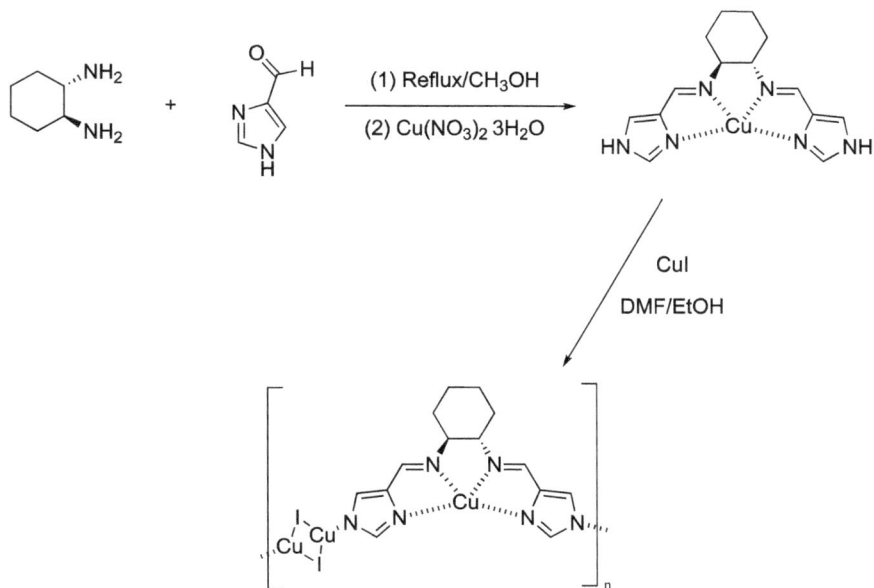

Figure 2.21 Solvothermal synthesis of the copper-containing salen polymer.

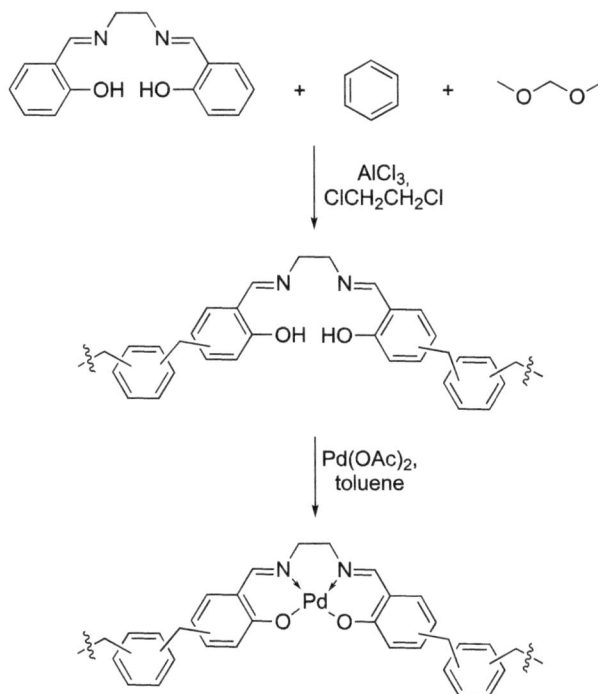

Figure 2.22 Post-polymerisation metallation of a salen polymer yields a palladium-containing organometallic polymer.

a coordination polymer (Figure 2.22). For example, advanced metal–aliphatic salen polymers with catalytic properties can be obtained by exploiting the susceptibility of the salen ligand to Friedel–Crafts alkylation[16] or pyrrole condensation (Figure 2.23).[17]

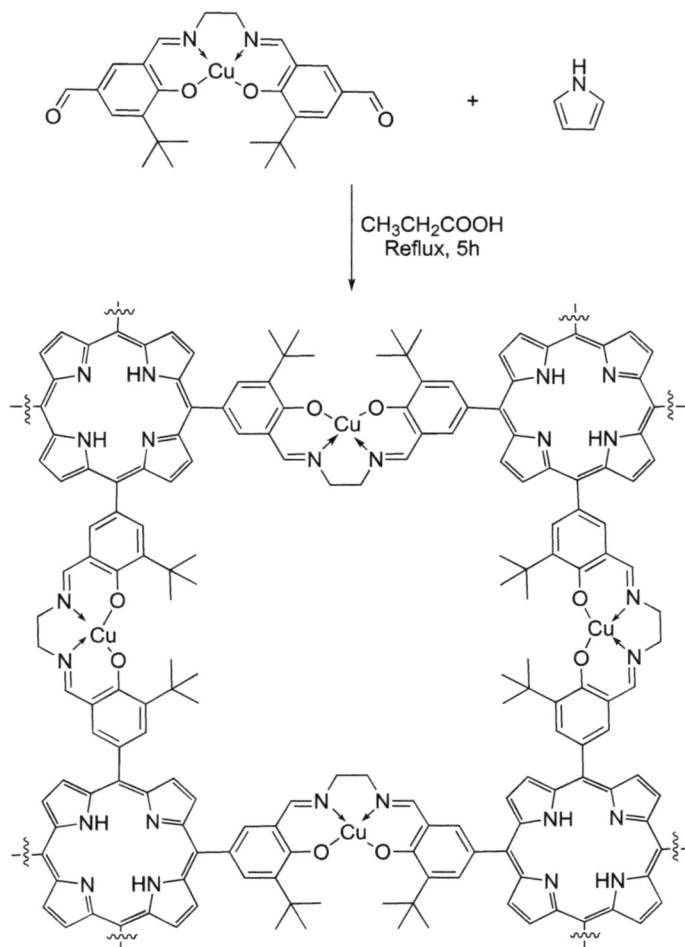

Figure 2.23 Condensation reaction between a salen complex and a pyrrole affords a copper-containing polymer.

2.4 Synthesis of Specific Classes of Polymers

Step polymerisation has been widely used to design different types of inorganic and organo-metallic polymers. The following subsections discuss these various types of polymers.

2.4.1 Synthesis of Polyphosphazenes

The polycondensation reaction of organophosphoranimines produces many high molecular weight polyphosphazenes. These reactions proceed *via* a chain mechanism but in a stepwise fashion. Phosphorus pentachloride, for instance, reacts with ammonia stepwise to produce a linear polymer alongside the hydrogen chloride byproduct (Figure 2.24) as well as cyclic products, such as cyclic chlorophosphazene (Figure 2.25). A breakthrough in the synthesis of polyphosphazenes *via* a step polymerisation route occurs after Neilson and Wisian-Neilson reported the high-temperature (180–200 °C)

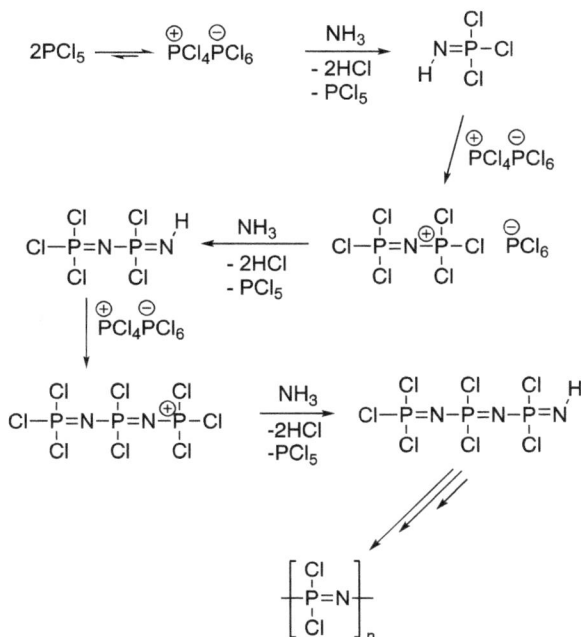

Figure 2.24 Stepwise condensation reaction of phosphorus pentachloride and ammonia yields a polyphosphazene.

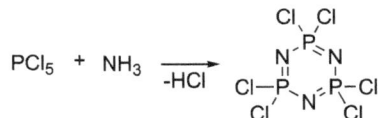

Figure 2.25 Cyclisation occurs during the stepwise condensation reaction of phosphorus pentachloride and ammonia, limiting the yield of the polyphosphazene.

polymerisation of XRR′P=NSiMe$_3$-type organo-*N*-silylphosphoranimines (Figure 2.26).[18] In the Neilson and Wisian-Neilson method, the macrostructure of the polymer depends on the leaving group (X) in the XRR′P=NSiMe$_3$-type organo-*N*-silylphosphoranimines. Linear polyorganophosphoranimines of about 50 000 Da are obtained when the leaving group is OCH$_2$CF$_3$, whereas a cyclic oligomeric tetramer results if the leaving group is a bromo group (Figure 2.26). The Neilson and Wisian-Neilson method has several drawbacks: (i) the method uses elevated temperature, which prevents successful polymerisation of monomers bearing reactive functional groups on the side chain (R and R′ substituents); (ii) the rates of polymerisation are slow and the polymer yields are moderate; (iii) there is a lack of control of molecular weight and molecular weight distribution.

Allcock and coworkers pioneered the ambient temperature step polymerisation method to synthesise polyphosphazenes. The Allcock method uses phosphorus pentachloride to initiate the cationic living chain polycondensation of trichloro(*N*-silyl)phosphoranimine (Figures 2.27 and 2.28).[19] This method produces high molecular weight (*ca.* 200 000 Da) polyphosphazenes with narrow molecular weight distribution ($Đ = 1.04$) in high yield and allows some control of the molecular weight by changing the monomer to initiator ratio.

(a)

$$\begin{array}{c} CH_3 \quad R' \\ H_3C\text{-}Si\text{-}N\text{=}P\text{-}X \\ CH_3 \quad R \end{array}$$

X is a leaving group

(b)

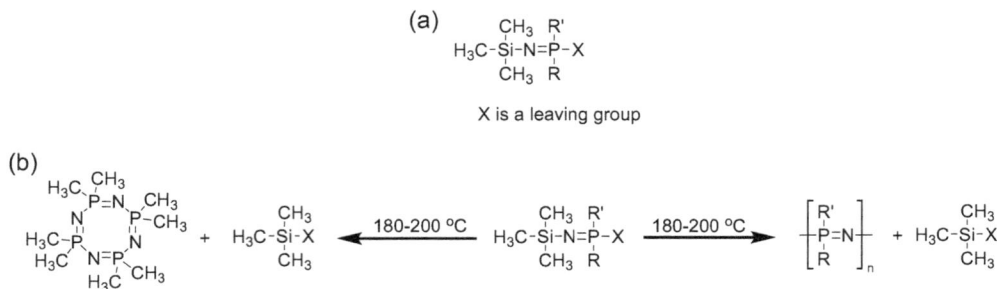

Figure 2.26 Neilson and Wisian-Neilson method for the synthesis of polyorganophosphazenes *via* condensation reaction of the XRR′P=NSiMe₃-type organo-*N*-silylphosphoran-imines (a). A linear polymer forms if X is OCH₂CF₃, while a cyclic oligomer results if X is a bromo group (Br) (b).

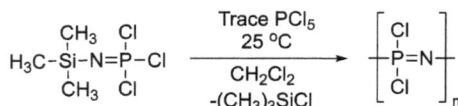

Figure 2.27   Ambient temperature synthesis of polyphosphazenes *via* step polymerisation.

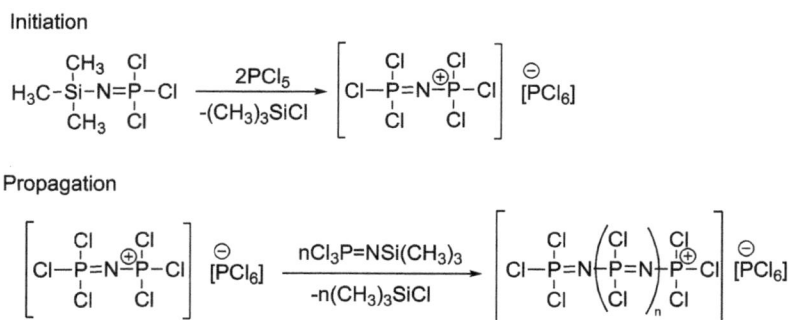

Initiation

Propagation

Figure 2.28   Polyphosphazenes are synthesised *via* a living chain cationic polycondensation involving initiation and propagation steps.

The method yields block copolymers by sequential addition of monomers and produces polyphosphazenes bearing reactive functional groups such as alkyne groups.

Other initiators for ambient temperature cationic living chain polycondensation of phosphoranimines include phosphites. Trimethyl phosphite initiates the polymerisation of *N*-silyl(halogeno)organophosphoranimines in chlorinated solvents to form high molecular weight homopolymers and copolymers of poly(alkyl/arylphosphazene) within 18 hours (Figure 2.29).[20] The successful synthesis of a linear polyphosphazene *via* a phosphite-

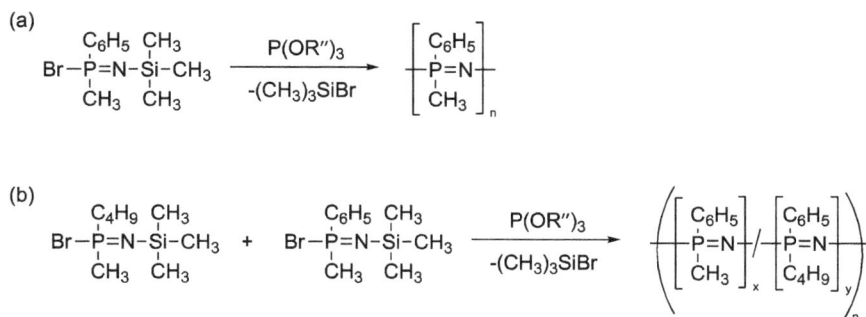

Figure 2.29 Phosphite-initiated living cationic polycondensation produces polyphosphazene homo- (a) and copolymers (b) at ambient temperature.

initiated polycondensation reaction of *N*-silyl(bromo)organophosphoranimine (Figure 2.29) contrasts with the formation of cyclic oligomers when this monomer was used in the Neilson and Wisian-Neilson high-temperature polycondensation method (Figure 2.26c). The polymerisation rate depends on the halide group, with chlorophosphoranimines polymerising more slowly than their bromo congener.[20] For phosphorus pentachloride-initiated polycondensation, the initiation step involves the reaction of two molecules of the initiator with one molecule of chlorophosphoranimine to yield a cationic initiator, which associates with a counteranion to form $[Cl_3P=N=PCl_3]^+[PCl_6]^-$ (Figure 2.28). The cationic initiator reacts with more monomers in the propagation step to give a living polymer alongside the trimethylsilyl chloride byproduct (Figure 2.28).[19] The counteranion has a substantial effect on the polymerisation rate. For example, with the phosphorus hexachloride anion, the polymerisation goes to completion in 4–6 hours, whereas with a weakly coordinating counteranion such as the tetrakis(pentafluorophenyl)borate anion, no polymerisation occurs in 24 hours.[21]

Anionic polycondensation of phosphoranimines gives polyphosphazene homo- and copolymers as well (Figure 2.30). The proposed mechanism of anionic polycondensation involves a tetra-*N*-butylammonium fluoride initiator, which abstracts the silyl group from the monomer to form a reactive anionic initiator after eliminating trimethylsilyl fluoride. Although this method gives high molecular weight (200 000 Da) homopolymers and copolymers, it has the disadvantage of being carried out at elevated temperature (130 °C) (Box 2.6).[22]

2.4.2 Synthesis of Polysilanes

Schlenk and Renning, and Kipping and Sands independently pioneered the dehalogenation of diorganodichlorosilanes, a reductive coupling of aryl- or alkyl-substituted silicon atoms to form polysilanes. Kipping and Sands synthesised several poly(diphenylsilane)s by heating two molar equivalents of sodium metal and diphenyldichlorosilane in bulk, toluene, or xylene.[23] C. A. Burkhard of the General Electric Co. Laboratory synthesised the first poly(dimethylsilane) by heating a mixture of sodium metal and dihalodimethylsilane in an autoclave.[24] Since the reaction only occurs with molten sodium, it is necessary to use high boiling point dihalodiorganosilanes such as dibromodimethylsilane (b.pt = 144 °C) or dichlorodimethylsilane (b.pt = 70 °C) or carry out the reaction in a sealed autoclave at elevated temperature and pressure. Burkhard's dehalogenation of dichlorodiorganosilanes (Figure 2.31) remains the most used method to synthesise high molecular weight homo- and

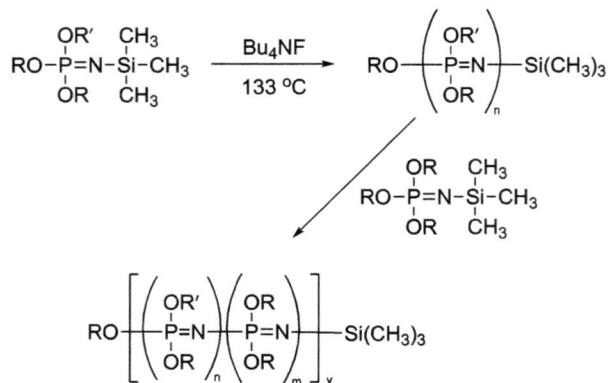

Figure 2.30 Anionic polycondensation of phosphoranimines gives polyphosphazene homo- and copolymers.

Box 2.6   Worked example 2.3.

Question

Propose the mechanism of the initiation and propagation steps of the anionic polycondensation of phosphoranimines.

Answer

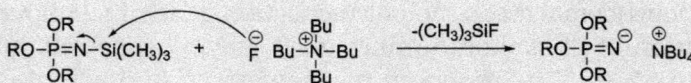

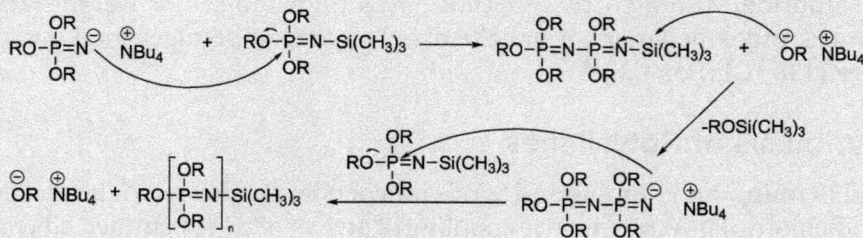

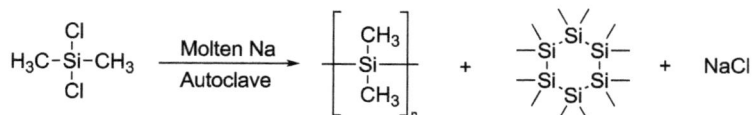

Figure 2.31 Burkhard's dehalogenation method gives a high molecular weight polysilane with a cyclic hexamer.

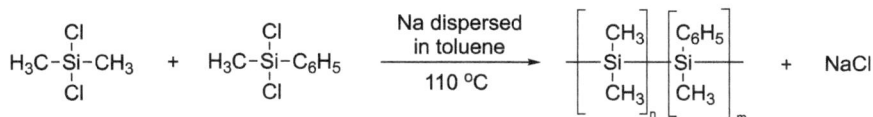

Figure 2.32 Wurtz reductive coupling reaction used in the synthesis of polysilanes.

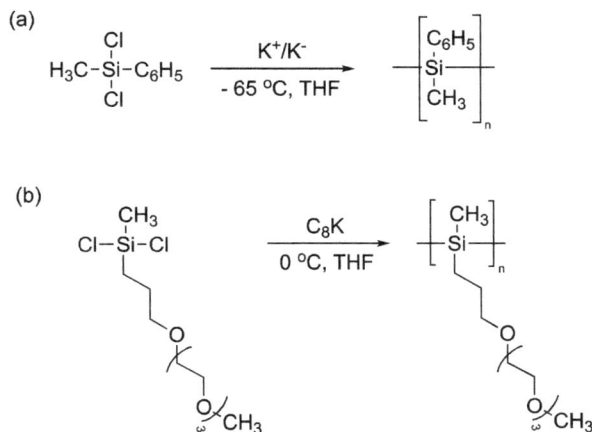

Figure 2.33 Modified Wurtz coupling reactions for low-temperature synthesis of polysilanes.

copolymers of polysilanes. A modern version of Burkhard's dehalogenation, a replica of Wurtz reductive coupling reaction, uses elevated temperature and dispersion of finely divided sodium in inert solvents, including toluene, xylene, ethers, or decane. The dichlorodiorganosilane monomer is added to the dispersion, or the sodium dispersion is added to the diorganodichlorosilane monomer. The polymerisation is a condensation reaction that eliminates the sodium chloride byproduct, but the reaction mechanism is reminiscent of a chain reaction (discussed in Chapter 3). A multimodal molecular weight distribution, consisting of cyclic, low molecular weight (~ 1000 Da) and high molecular weight ($> 100\,000$ Da) polymers, results during the polymerisation. The multimodal distribution is probably due to multiple reaction mechanisms controlling the chain propagation step or heterogeneous nature of the reaction system, for instance, the use of sodium dispersion. Cyclic polymers are the exclusive products at equilibrium, while high molecular weight polymers predominate under kinetic control. A single dichlorodiorganosilane monomer results in a homopolymer, while a mixture of different dichlorodiorganosilane monomers gives a copolymer (Figure 2.32).

In contrast to the Burkhard and Wurtz methods, a modified Wurtz method based on low temperature ($-65\,^{\circ}$C to $25\,^{\circ}$C) synthesis is feasible when a graphite-potassium (C_8K) or potassium ion-pair complex [K^+/K^-] serves as the reducing agent instead of sodium (Figure 2.33).[25,26] At low temperature, the reaction gives bimodal or trimodal molecular weight distribution, low yields ($< 50\%$), and low molecular weight polymers (~ 2000 Da). The addition of a crown ether to the reaction increases the polymerisation rate and changes the molecular weight distribution from bimodal to monomodal. The crown ether reacts with alkaline metals such as potassium and sodium to form a crown ether–sodium ion-pair complex, which increases the polymerisation rate.

Sonochemical polymerisation using ultrasonic agitation of the reaction mixture also lowers the polymerisation temperature, allowing the synthesis of polysilanes with monomodal molecular weight distribution and near-uniform dispersity at low temperature. Sonochemical homopolymerisation at ambient temperature is successful with diaryldichlorosilane monomers in nonpolar aromatic solvents such as toluene and xylene but unsuccessful with dialkyldichlorosilane monomers. Under these conditions, dialkyldichlorosilane undergoes copolymerisation with phenylmethyldichlorosilane. With ultrasound irradiation and ethereal solvents, dialkyldichlorosilane monomers, exemplified by dihexyldichlorodisilane, undergo reductive coupling to give homopolymers such as poly(dihexylsilane). The use of microwave heating also increases the polymerisation rate and improves the molecular weight distribution of the polymer.

2.4.3   Synthesis of Polysiloxanes

More popularly known as silicones, polysiloxanes are undoubtedly one of the oldest and most commercialised inorganic polymers. Silly putty, known for its unusual flow properties, exemplifies a polysiloxane-based product, where the characteristic flow is due to polydimethylsiloxane. Ring-opening polymerisation (discussed in Chapter 3) of cyclic siloxanes obtained *via* hydrolysis of an appropriate dihalodialkylsilane is the preeminent method to synthesise polysiloxane homopolymers and random copolymers. However, polysiloxanes can be prepared by the condensation reaction of water with an appropriate organosilicon such as dihalodialkylsilanes. Synthesis of polysiloxanes *via* the polycondensation route (Figure 2.34) becomes convenient and classic after F. S. Kipping applied the Grignard method to synthesise dihalodialkylsilane monomers (Figure 2.34a).[27] The Grignard method has been replaced by the Rochow process which synthesises dihalodiorganosilanes from elemental silicon (Figure 2.34b). Overall, the polycondensation route remains the method of choice to synthesise well-defined and special polysiloxane copolymers such as silphenylene-containing polysiloxanes (Figure 2.35).[28]

Different versions of the polycondensation method now exist. Merker's method involving the condensation reaction of 1,4-bis(dimethylhydroxysilyl)benzene produces poly[(oxydimethylsilylene)(1,4-phenylene)(dimethylsilylene)] in low yields while requiring elevated temperatures and tedious recrystallisation (Figure 2.36a).[29] Another polycondensation method is the transition metal-catalysed cross-dehydrocoupling reaction of an organosilane with water to produce a high molecular weight poly(carbosiloxane) in high yield (Figure 2.36b). This single step cross-dehydrocoupling polycondensation reaction (Figure 2.36b) is attractive because it occurs at ambient

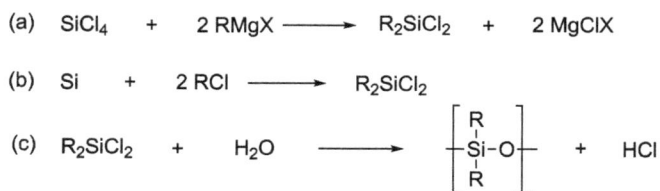

$$\text{(a)} \quad SiCl_4 \quad + \quad 2\,RMgX \longrightarrow R_2SiCl_2 \quad + \quad 2\,MgClX$$

$$\text{(b)} \quad Si \quad + \quad 2\,RCl \longrightarrow R_2SiCl_2$$

$$\text{(c)} \quad R_2SiCl_2 \quad + \quad H_2O \longrightarrow \left[\begin{matrix} R \\ | \\ Si-O \\ | \\ R \end{matrix} \right]_n \quad + \quad HCl$$

Figure 2.34   Synthesis of dihalodiorganosilanes *via* (a) Grignard and (b) Rochow processes. (c) Hydrolysis of dihalodiorganosilanes yields polysiloxanes.

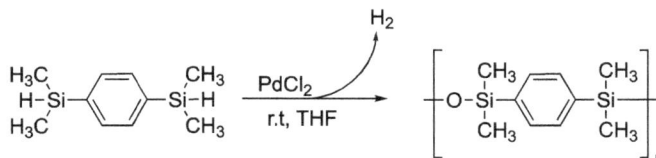

Figure 2.35   Cross-dehydrocoupling polycondensation in the synthesis of silphenylene-containing polysiloxanes.

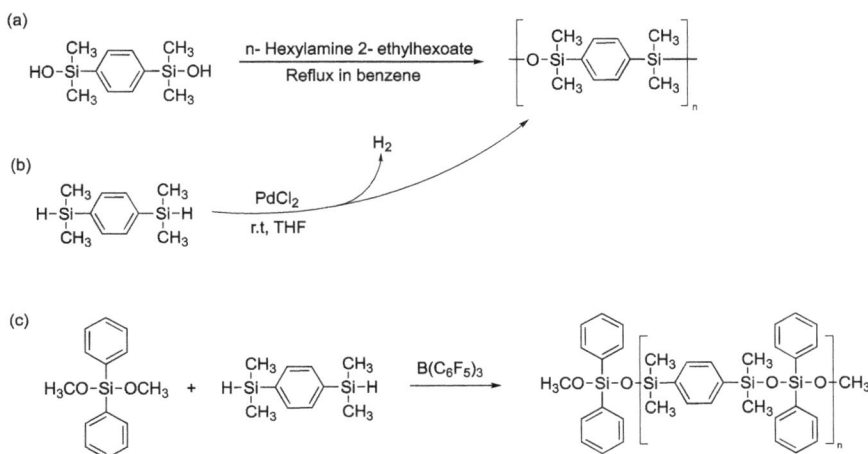

Figure 2.36   Synthesis of polysiloxanes *via* (a) Merker's process, (b) ambient metal-catalysed cross-dehydrocoupling of organohydrosilanes, and (c) ambient dehydrogenative coupling of organohydrosilanes.

temperature and produces the hydrogen gas byproduct, which is easily removable from the reaction vessel.[28] Also, transition metal-catalysed dehydrogenative coupling of organohydrosilanes such as 1,4-bis(dimethylsilyl)benzene with organosilanols such as disilanol provides a convenient and highly selective route to silphenylene-containing polysiloxanes at room temperature.[30] The Lewis acid, tris(pentafluorophenyl)borane, also catalyses the dehydrogenative coupling of organohydrosilanes and organoalkosilanes at room temperature or below to yield the same silphenylene–siloxane copolymer (Figure 2.36c).[31] The copolymer is unique because it combines the flexibility and thermal stability (discussed in Chapter 5) of the siloxane linkage with the rigid aromatic backbone in the main polymer chain (Box 2.7).

2.4.4   Synthesis of Polystannanes and Polygermanes

If carbon and silicon form polymers, can other group IV elements, such as tin and germanium, form polymers? Indeed, polymers of tin, known as polystannanes, and germanium, known as polygermanes, exist. Polystannanes are particularly unique, being the only known polymers having the main chain derived entirely from metal atoms, and these polymers, including polygermanes, are being investigated for their electrical properties. Like the reductive coupling method used to synthesise polysilanes, the Wurtz coupling reaction of dialkyldichlorostannanes or dialkyldichlorogermanes using sodium in toluene yields polystannanes or polygermanes (Figure 2.37), respectively.[32–34]

Box 2.7 Worked example 2.4.

Question

Starting with silicone and assuming all reagents are commercially available, provide a synthetic route to poly(dimethylsilane)-*co*-(dihexylsilane).

Answer

Synthesise dichlorodimethylsilane *via* the Rochow process

$$Si \ + \ CH_3Cl \ \longrightarrow \ CH_3SiCl_2$$

Synthesise dichlorodihexylsilane *via* the Rochow process

$$Si \ + \ C_6H_{13}Cl \ \longrightarrow \ C_6H_{13}SiCl_2$$

Synthesise copolymer via Wurtz reductive coupling of dichlorodimethylsilane and dichlorodihexylsilane

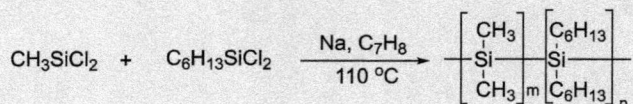

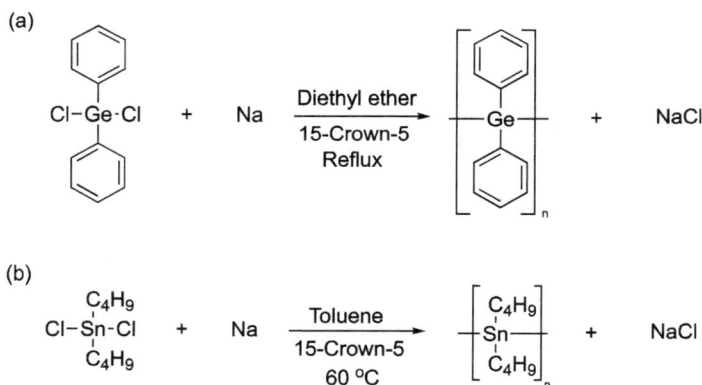

Figure 2.37 The Wurtz coupling reaction yields a (a) polygermane and (b) polystannane from the reaction of sodium with dichlorodiorganogermane and dichlorodiorganostannane.

For example, the Wurtz coupling reaction produces an insoluble polygermane when dichlorodiphenylgermane was refluxed with sodium in diethyl ether in the presence of 15-crown-5. Similarly, the Wurtz coupling reaction produces high molecular weight polystannanes from the reaction of dibutyldichlorostannane with sodium in the presence of 15-crown-5.

Dialkyldichlorostannane or dialkyldichlorogermane monomers can also undergo electrochemical polymerisation to yield the corresponding polystannanes or polygermanes, respectively.[35–37] Electrochemical reduction of butyltrichloro-stannane in 1,2-dimethoxyethane containing tetrabutylammonium perchlorate as the supporting electrolyte and a platinum plate and silver wire as the cathode and anode, respectively, resulted in a polystannane network polymer (Figure 2.38a). In contrast, a linear

(a)

$$Cl\text{-}\underset{\underset{Cl}{|}}{\overset{\overset{C_4H_9}{|}}{Sn}}\text{-}Cl \xrightarrow[\underset{CH_3OCH_2CH_2OCH_3}{(C_4H_9)_4N(ClO_4)\ in}]{e^-} [C_4H_6Cl_xSn]_n \quad + \quad Cl^-$$

$$x = 0.2$$

(b)

$$Cl\text{-}\underset{\underset{C_6H_5}{|}}{\overset{\overset{C_4H_9}{|}}{Ge}}\text{-}Cl \xrightarrow[LiClO_4\ in\ THF]{e^-} \left[\!\!\begin{array}{c} C_4H_9 \\ | \\ Ge \\ | \\ C_6H_5 \end{array}\!\!\right]_n \quad + \quad Cl^-$$

(c)

$$Cl\text{-}\underset{\underset{C_6H_5}{|}}{\overset{\overset{C_4H_9}{|}}{Ge}}\text{-}Cl \quad + \quad Cl\text{-}\underset{\underset{C_6H_5}{|}}{\overset{\overset{CH_3}{|}}{Si}}\text{-}Cl \xrightarrow[LiClO_4\ in\ THF]{e^-} \left[\!\!\begin{array}{c} C_4H_9 \\ | \\ Ge \\ | \\ C_6H_5 \end{array}\!\!\right]_n\left[\!\!\begin{array}{c} CH_3 \\ | \\ Si \\ | \\ C_6H_5 \end{array}\!\!\right]_m \quad + \quad Cl^-$$

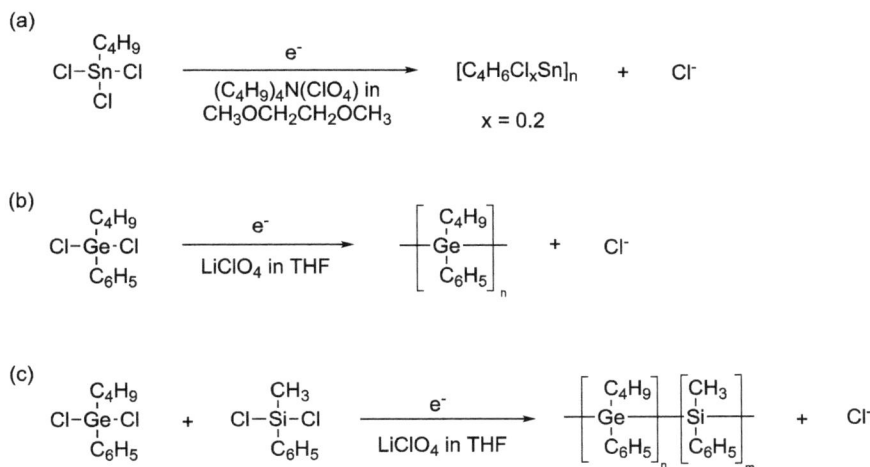

Figure 2.38 Electrochemical polymerisation yields a (a) polystannane, (b) polygermane, and (c) polygermane-*co*-polysilane.

Box 2.8 Worked example 2.5.

Question

Use chemical equations to present the strategy for the synthesis of a poly-(butylphenylsilane)-*co*-poly(dibutylstannane)-*co*-poly(diethylgermane). Assume all monomers are commercially available.

Answer

Wurtz coupling strategy

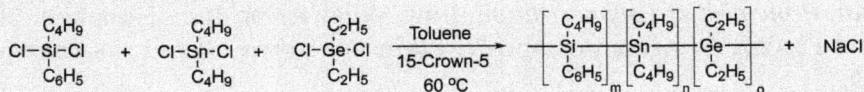

$$Cl\text{-}\underset{\underset{C_6H_5}{|}}{\overset{\overset{C_4H_9}{|}}{Si}}\text{-}Cl + Cl\text{-}\underset{\underset{C_4H_9}{|}}{\overset{\overset{C_4H_9}{|}}{Sn}}\text{-}Cl + Cl\text{-}\underset{\underset{C_2H_5}{|}}{\overset{\overset{C_2H_5}{|}}{Ge}}\text{-}Cl \xrightarrow[\substack{15\text{-Crown-5}\\60\ ^\circ C}]{Toluene} \left[\!\!\begin{array}{c} C_4H_9 \\ | \\ Si \\ | \\ C_6H_5 \end{array}\!\!\right]_m\left[\!\!\begin{array}{c} C_4H_9 \\ | \\ Sn \\ | \\ C_4H_9 \end{array}\!\!\right]_n\left[\!\!\begin{array}{c} C_2H_5 \\ | \\ Ge \\ | \\ C_2H_5 \end{array}\!\!\right]_o + NaCl$$

Electropolymerisation strategy

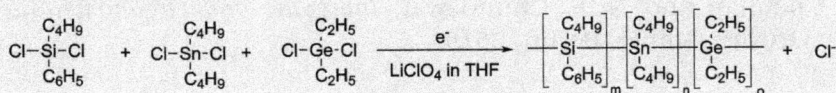

$$Cl\text{-}\underset{\underset{C_6H_5}{|}}{\overset{\overset{C_4H_9}{|}}{Si}}\text{-}Cl + Cl\text{-}\underset{\underset{C_4H_9}{|}}{\overset{\overset{C_4H_9}{|}}{Sn}}\text{-}Cl + Cl\text{-}\underset{\underset{C_2H_5}{|}}{\overset{\overset{C_2H_5}{|}}{Ge}}\text{-}Cl \xrightarrow[LiClO_4\ in\ THF]{e^-} \left[\!\!\begin{array}{c} C_4H_9 \\ | \\ Si \\ | \\ C_6H_5 \end{array}\!\!\right]_m\left[\!\!\begin{array}{c} C_4H_9 \\ | \\ Sn \\ | \\ C_4H_9 \end{array}\!\!\right]_n\left[\!\!\begin{array}{c} C_2H_5 \\ | \\ Ge \\ | \\ C_2H_5 \end{array}\!\!\right]_o + Cl^-$$

---

polygermane having a molecular weight of $M_n = 19\,000$ Da is obtained *via* the electroreduction of dichlorobutylphenylgermane in tetrahydrofuran containing lithium perchlorate as the supporting electrolyte and magnesium as the electrode (Figure 2.38b). Similar electrochemical conditions yield polygermane-*co*-polysilane from a mixture of dichlorobutylphenylgermane and dichloromethylphenylsilane (Figure 2.38c). Another method to synthesise polystannanes and polygermanes is the catalytic dehydropolymerisation of dialkyldihydrostannanes or dialkyldihydrogermanes, respectively (Boxes 2.8 and 2.9). For example, dibutylstannane undergoes $[RhCl(PPh_3)_3]$-catalysed

Box 2.9 Review questions.

1. Describe how to design a cobaltocene-containing polymer, wherein the cobalt is in the main chain of the polymer. Use equations.
2. Describe how to synthesise a block copolymer of platinum- and iridium-containing polymetallaynes incorporating platinum and iridium in the main chain.
3. Describe how to synthesise a block copolymer containing poly(ethylphenylsilane) and poly(methylphenylsilane) blocks.

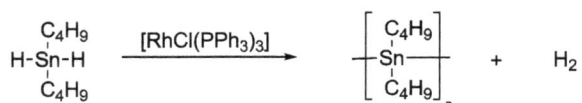

Figure 2.39 Catalytic dehydropolymerisation of dialkyldihydrostannane results in a polystannane with the elimination of hydrogen gas.

dehydropolymerisation to give linear poly(butylstannane) and the hydrogen gas by-product without forming low molecular weight cyclic polymers, which are commonly produced in Wurtz coupling polymerisation (Figure 2.39).[38]

Further Reading

1. J. E. Mark, H. R. Allcock and R. West, *Inorganic Polymers*, Oxford University Press Inc., New York, 2005.
2. R. D. Archer, *Inorganic and Organometallic Polymers*, Wiley-VCH, New York, 2001.
3. G. Odian, *Principles of Polymerization*, John Wiley & Son Inc., Hoboken, 2004.
4. A. S. Abd-El-Aziz, C. Agatemor and N. Etkin, *Macromol. Rapid Commun.*, 2014, **35**, 513–559.
5. C. U. Pittman Jr, C. E. Carraher Jr, M. Zeldin, J. E. Sheets and B. M. Culbertson, *Metal-containing Polymeric Materials*, Plenum Press, New York, 1996.
6. N. P. S. Chauhan and N. S. Chundawat, *Inorganic and Organometallic Polymers*, Walter de Gruyter GmbH, Berlin, 2019.

References

1. G. Odian, *Principles of Polymerization*, Hoboken, NJ, John Wiley & Sons, 2004.
2. H. K. Kim and K. Matyjaszewski, *J. Polym. Sci., Part A: Polym. Chem.*, 1993, **31**, 299–307.
3. R. D. Archer, *Inorganic and Organometallic Polymers*, New York, John Wiley & Son, 2001.
4. A. S. Abd-El-Aziz, C. Agatemor and N. Etkin, *Macromol. Rapid Commun.*, 2014, **35**, 513–559.
5. A. S. Abd-El-Aziz, J. L. Pilfold, B. Z. Momeni, A. J. Proud and J. K. Pearson, *Polym. Chem.*, 2014, **5**, 3453–3465.
6. M. Altmann and U. H. Bunz, *Angew. Chem., Int. Ed.*, 1995, **34**, 569–571.
7. A. S. Abd-El-Aziz, C. Agatemor, W.-Y. Wong, *Macromolecules Incorporating Transition Metals: Tackling Global Challenges*, Royal Society Publishing, Cambridge, UK, 2018.
8. W.-Y. Wong, X. Wang, H.-L. Zhang, K.-Y. Cheung, M.-K. Fung, A. B. Djurišić and W.-K. Chan, *J. Organomet. Chem.*, 2008, **693**, 3603–3612.

9. W.-Y. Wong, X.-Z. Wang, Z. He, A. B. Djurišić, C.-T. Yip, K.-Y. Cheung, H. Wang, C. S. Mak and W.-K. Chan, *Nat. Mater.*, 2007, **6**, 521–527.
10. W. Steffen, B. Köhler, M. Altmann, U. Scherf, K. Stitzer, H. zur Loye and U. H. Bunz, *Chem. – Eur. J.*, 2001, **7**, 117–126.
11. G. L. Schulz and S. Holdcroft, *Chem. Mater.*, 2008, **20**, 5351–5355.
12. T. A. Clem, D. F. Kavulak, E. J. Westling and J. M. Fréchet, *Chem. Mater.*, 2010, **22**, 1977–1987.
13. I. W. Shim and W. M. Risen Jr, *J. Organomet. Chem.*, 1984, **260**, 171–179.
14. M. R. Roner, C. E. Carraher, K. Shahi, Y. Ashida and G. Barot, *BMC Cancer*, 2009, **9**, 358.
15. Y.-L. Hou, R. W.-Y. Sun, X.-P. Zhou, J.-H. Wang and D. Li, *Chem. Commun.*, 2014, **50**, 2295–2297.
16. D. Meng, J. Bi, Y. Dong, B. Hao, K. Qin, T. Li and D. Zhu, *Chem. Commun.*, 2020, **56**, 2889–2892.
17. X. Sun, F. Meng, Q. Su, K. Luo, P. Ju, Z. Liu, X. Li, G. Li and Q. Wu, *Dalton Trans.*, 2020, **49**, 13582–13587.
18. P. Wisian-Neilson and R. H. Neilson, *J. Am. Chem. Soc.*, 1980, **102**, 2848–2849.
19. V. Blackstone, A. J. Lough, M. Murray and I. Manners, *J. Am. Chem. Soc.*, 2009, **131**, 3658–3667.
20. T. J. Taylor, A. Presa Soto, K. Huynh, A. J. Lough, A. C. Swain, N. C. Norman, C. A. Russell and I. Manners, *Macromolecules*, 2010, **43**, 7446–7452.
21. V. Blackstone, S. Pfirrmann, H. Helten, A. Staubitz, A. Presa Soto, G. R. Whittell and I. Manners, *J. Am. Chem. Soc.*, 2012, **134**, 15293–15296.
22. R. A. Montague and K. Matyjaszewski, *J. Am. Chem. Soc.*, 1990, **112**, 6721–6723.
23. F. S. Kipping and J. E. Sands, *J. Chem. Soc., Trans.*, 1921, **119**, 830–847.
24. C. A. Burkhard, *J. Am. Chem. Soc.*, 1949, **71**, 963–964.
25. B. Lacave-Goffin, L. Hevesi and J. Devaux, *J. Chem. Soc., Chem. Commun.*, 1995, 769–770.
26. B. Lacave-Goffin, L. Hevesi and J. Devaux, *Chem. Commun.*, 1996, 765–766.
27. F. S. Kipping and L. L. Lloyd, *J. Chem. Soc., Trans.*, 1901, **79**, 449–459.
28. Y. Li and Y. Kawakami, *Macromolecules*, 1999, **32**, 3540–3542.
29. R. L. Merker and M. J. Scott, *J. Polym. Sci., Part A: Gen. Pap.*, 1964, **2**, 15–29.
30. Y. Li and Y. Kawakami, *Macromolecules*, 1999, **32**, 8768–8773.
31. S. Rubinsztajn and J. A. Cella, *Macromolecules*, 2005, **38**, 1061–1063.
32. D. Miles, T. Burrow, A. Lough and D. Foucher, *J. Inorg. Organomet. Polym. Mater.*, 2010, **20**, 544–553.
33. N. Devylder, M. Hill, K. C. Molloy and G. J. Price, *Chem. Commun.*, 1996, 711–712.
34. R. H. Cragg, R. G. Jones, A. C. Swain and S. J. Webb, *J. Chem. Soc., Chem. Commun.*, 1990, 1147–1148.
35. S. Aeiyach, P.-C. Lacaze, J. Satgé and G. Rima, *Synth. Met.*, 1993, **58**, 267–270.
36. M. Okano, K. Watanabe and S. Totsuka, *Electrochemistry*, 2003, **71**, 257–259.
37. T. Shono, S. Kashimura and H. Murase, *J. Chem. Soc., Chem. Commun.*, 1992, 896–897.
38. F. Choffat, P. Smith and W. Caseri, *J. Mater. Chem.*, 2005, **15**, 1789–1792.

3 Chain Polymerisation

3.1 Introduction

This chapter focuses on the chain polymerisation of inorganic and organometallic monomers. This polymerisation technique is widely employed to synthesise high molecular weight polymers and, by optimising the polymerisation conditions, to produce well-defined polymers with controlled molecular weight and dispersity. Depending on the mechanism, the polymerisation can proceed *via* a cationic, an anionic, or a radical route, providing options to accommodate diverse monomers and reaction conditions. This versatility allows the use of chain polymerisation to design organometallic polymers of varied microstructures, macrostructures, and functionalities. Unlike step polymerisation (discussed in Chapter 2) which utilises monomers bearing functional groups, chain polymerisation employs unsaturated monomers, particularly alkenes. Usually, an organometallic moiety, such as a metallocene, is conjugated to an unsaturated monomer such as an acrylate to obtain chain polymerisable monomers. In Section 2.1 of Chapter 2, we mentioned some dilemmas faced in synthesising and polymerising organometallic monomers. These dilemmas are also encountered in chain polymerisation. This chapter introduces the synthesis strategies used in the chain polymerisation of inorganic and organometallic monomers. We will discuss the fundamental principles of chain polymerisation to guide the student in using this technique to design inorganic and organometallic polymers (Box 3.1).

Box 3.1 Learning outcomes.

By the end of this chapter, the student should be able to:

1. Differentiate between step and chain polymerisations.
2. Describe the thermodynamic and kinetic factors that influence chain polymerisation.
3. Predict the effect of inorganic and organometallic moieties on the efficiency of chain polymerisation.

Fundamentals of Inorganic and Organometallic Polymer Science
By Christian Agatemor, Kajal Ghosal, Samuel Fura and Peter J. S. Foot
© Christian Agatemor, Kajal Ghosal, Samuel Fura and Peter J. S. Foot 2024
Published by the Royal Society of Chemistry, www.rsc.org

4. Propose mechanisms for the chain polymerisation of inorganic and organo-metallic monomers.
5. Design inorganic and organometallic polymers using chain polymerisation.
6. Differentiate between "controlled" and "non-controlled" chain polymerisations.
7. Design polymerisation conditions to achieve controlled polymerisation.
8. Describe ring-opening polymerisation and ring-opening olefin metathesis polymerisation of inorganic and organometallic monomers.
9. Identify the conditions that favour strain-driven and entropy-driven ring-opening polymerisations.

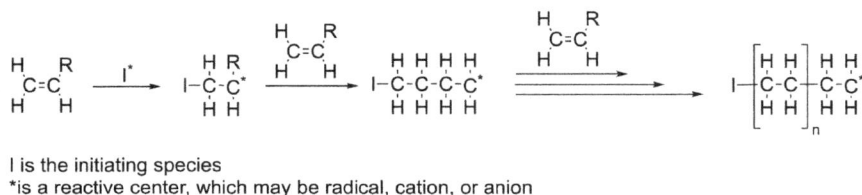

I is the initiating species
*is a reactive center, which may be radical, cation, or anion

Figure 3.1 Schematic representation of the polymerisation of a carbon–carbon double bond.

3.2 Fundamental Aspects of Chain Polymerisation

In Chapter 2, we discussed how to use step polymerisation to synthesise inorganic and organometallic polymers. The following section will consider chain polymerisation, which is also used to synthesise many of these polymers. The mechanism of step polymerisation is distinct from that of chain polymerisation. For example, step polymerisation yields high molecular weight polymers only at high monomer conversion (>98%) in contrast to chain polymerisation which produces high molecular weight polymers at all monomer conversions (also discussed in Chapter 1). At any time, a typical chain polymerisation reaction vessel lacks intermediate-sized polymers, containing only monomers, large-sized polymers, and initiator molecules. The initiator molecule generates reactive species – radicals, cations, or anions – to initiate the polymerisation. Monomers in chain polymerisation must contain an unsaturated functional group, usually a pi bond, that reacts with the reactive species to generate a growing polymer that bears a reactive centre (Figure 3.1). The alkene double bond is the most polymerisable functional group and is amenable to radical, cationic, and anionic radical polymerisations. The polarised nature of polar functional groups such as carbonyl that contain a pi bond makes them non-susceptible to radical polymerisation; however, these groups undergo cationic and anionic polymerisations. Chain polymerisation proceeds by propagating the reactive centres as monomers are successively added to elongate the growing polymer chain (Figure 3.1). Termination of polymerisation occurs by deactivation of the reactive centre by an appropriate reaction. Chain polymerisation, therefore, consists of different reaction steps: chain initiation, chain propagation, chain termination and, in some instances, chain transfer. Depending on the initiator and the reactive species, chain polymerisation can be classified as radical, anionic, or cationic (Box 3.2).

Box 3.2 Reactions in chain polymerisation.

Chain initiation: this is the first step in chain polymerisation and generates reactive species from the initiating molecule. Chain initiation proceeds in two steps. First, the initiating molecule undergoes cleavage, generating reactive initiating species, radicals, cations, or anions. Second, the initiating species initiate the polymerisation by transferring their reactive centre to monomers.

Initiation steps of radical chain polymerisation

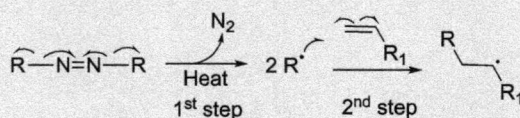

Initiation steps of cationic chain polymerisation

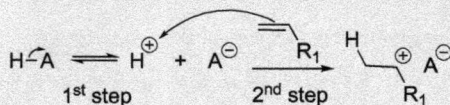

Initiation steps of anionic step polymerisation

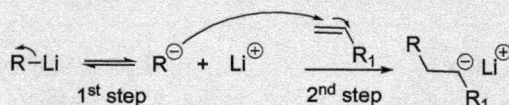

Chain propagation: once the reactive initiating species are generated, they add to monomers, forming reactive monomers. These reactive monomers add to new monomers, which acquire the reactive centre to further react with new monomers in a chain-like process. The process is iterative and propagates the chain until the monomers are depleted or the reaction is terminated. Chain propagation predominates the polymerisation process.

Propagation step illustrated with radical chain polymerisation

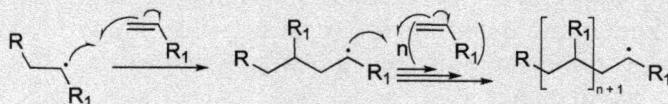

Chain transfer: this reaction transfers the reactivity of the growing polymer chain to another molecule that is incapable of propagating the polymerisation. In all chain transfer reactions, hydrogen or another molecule is abstracted from a solvent, an initiator, a monomer, a growing polymer, or a chain transfer agent, then transferred to another growing polymer chain. This reaction terminates the polymerisation and reduces the average molecular weight of the final polymer.

Chain transfer to a solvent molecule in a radical chain polymerisation

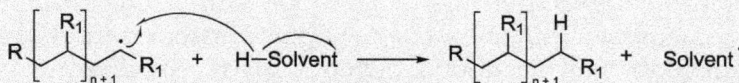

Chain termination: this reaction stops polymer chains from forming with a reactive centre, inhibiting the polymerisation. Depletion of monomers or the addition of a reactive impurity effectively terminates the entire polymerisation. In radical chain polymerisation, combination, disproportionation, and reaction with an initiating radical are common mechanisms to terminate the reaction.

Chain termination by reaction with oxygen impurity

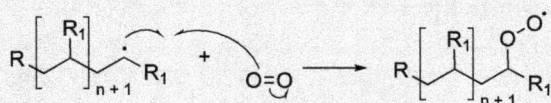

Chain termination by combination of two growing polymer chains

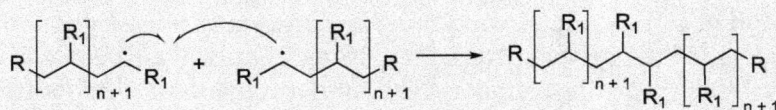

Chain termination by radical disproportionation

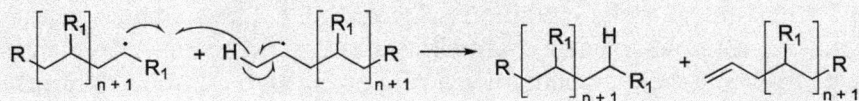

3.2.1 Thermodynamic and Kinetic Perspectives

Chain polymerisable monomers must contain an unsaturated bond. In addition, polymerisability depends on favourable thermodynamic and kinetic factors operating under the reaction conditions. A typical chain polymerisation is predominated by the chain propagation step, with relatively few initiation and termination steps; therefore, we will focus our consideration of the chemical thermodynamics only to the propagation step. Like other chemical reactions, the state functions of enthalpy (H), entropy (S) and Gibbs free energy (G) characterise the thermodynamics of the polymerisation. The quantities ΔH, ΔS, and ΔG define the differences in enthalpy, entropy, and Gibbs free energy between one mole of monomer and one mole of repeating unit in the resulting polymer. The state functions are related by eqn (3.1):

$$\Delta G = \Delta H - T\Delta S \tag{3.1}$$

Chain polymerisations are typically exothermic reactions (negative ΔH) because the propagation step transforms the unsaturated bond of the monomer into a saturated bond in the polymer, except in the case of ring-opening olefin metathesis polymerisation (discussed in Section 3.6). Orderliness increases as monomers link up to form polymers, resulting in an unfavourable entropy term (negative ΔS) for polymerisation. For most unsaturated monomers, polymerisation is thermodynamically feasible (negative ΔG) because the negative ΔH term contributes more to ΔG than to the $T\Delta S$ term in eqn (3.1). Polymerisation depends on thermodynamic feasibility and kinetic feasibility, the latter related to the experimental conditions that allow the polymerisation to proceed at a realistic rate. For unsaturated monomers containing inorganic or organometallic moieties, a thermodynamically feasible polymerisation will require special reaction conditions to ensure kinetic feasibility.

3.2.2   Impact of Organometallic Moieties on Polymerisation

Eqn (3.1) implies that changes in the enthalpy and entropy of polymerisation influence the reactivity of monomers at a given temperature. For unsaturated monomers, substituents that perturb the electron-cloud density at the pi bond, introduce steric strain or create electronic interactions affect the enthalpy of polymerisation. Substituents that introduce a resonance effect, an inductive effect or hyperconjugation can perturb the electron cloud density at the pi bond of the monomer to alter the polymerisation kinetics. For example, in the cationic chain polymerisation of olefins, the pi bond must be sufficiently nucleophilic to react with electrophilic initiators, or the growing cationic polymer chains must be stable enough to grow into high molecular weight polymers. Electron-rich organometallic molecules, such as cyclopentadienylmanganese tricarbonyl, ferrocene, ruthenocene, or osmocene, donate electrons to the pi bond, making the bond highly nucleophilic towards weakly electrophilic initiators such as hydrogen azide and acetic acid which are unreactive towards monosubstituted olefins under similar conditions (Figure 3.2).[1,2] These organometallic molecules stabilise a growing cationic polymer chain either through resonance effects by delocalising the positive charge over the cyclopentadienyl ring or through the molecular orbital overlap of the metal atom with the p-orbital of the positively charged α-carbon atom. For metallocenes, the order of reactivity of vinylmetallocenes toward weak electrophiles correlates with increasing ionic radius of the metal atom, implying that the metal contributes to the nucleophilicity of the pi bond. Vinyl monomers conjugated to organometallic molecules also undergo radical and anionic chain polymerisations, where the organometallic moiety affects the monomer reactivity and growing polymer stability. In all these polymerisations, the growing polymer is stabilised by electron or charge delocalisation over the cyclopentadienyl ligand or overlap of the metal orbital with the p-orbital of the carbon atom.

In addition to influencing the initiation and propagation steps, organometallic moieties affect the termination step as well. For example, the termination mechanism

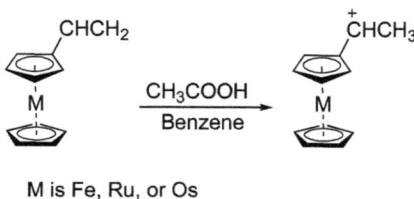

M is Fe, Ru, or Os

Figure 3.2   Electrophilic addition of a weak electrophile, acetic acid, to vinylmetallocenes. The polymerisation rate increases down the group (Fe < Ru < Os).

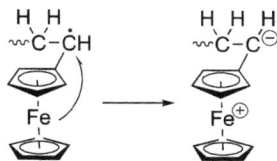

Figure 3.3 A mechanism of chain termination *via* intramolecular electron transfer from the organometallic moiety to the growing polymeric radical.

in the radical chain polymerisation of vinylferrocene is distinct from that of organic vinyl monomers such as styrene.[3] Usually, radical polymerisation is terminated by annihilating the radical species through intermolecular reactions (Box 3.2). However, in vinylferrocene, polymerisation can be terminated by intramolecular electron transfer from a ferrocene moiety to the growing polymeric radical (Figure 3.3). This intramolecular electron transfer annihilates the growing chain radical, producing a polymer with a high spin iron(III) complex (Figure 3.3).[3] While ferrocene-containing vinyl monomers feature a unique mechanism of termination, the mechanisms of the initiation and propagation steps are essentially similar to those of styrene under identical conditions.[4] This is not surprising considering that radicals are neutral, highly reactive and do not require stringent conditions to react with pi bonds.

Unlike the indiscriminate reactivity of radicals, cations and anions are highly selective. Nonetheless, conjugating an organometallic moiety to an olefinic monomer alters ionic polymerisation kinetics. The change in initiation and propagation rates, for instance, has an important implication for the design of copolymers. In an anionic chain copolymerisation of methyl methacrylate monomers to produce poly(ferrocenylmethyl methacrylate)-*block*-poly(methyl methacrylate), the methyl methacrylate block is first produced before the ferrocenylmethyl block (Figure 3.4) because the ferrocenylmethyl substituent stabilises the propagating carbanion more than methyl methacrylate.[5] The rationale is that a growing anionic poly(ferrocenylmethyl methacrylate) is less nucleophilic than a growing poly(methyl methacrylate), and therefore less reactive in anionic polymerisation. For the copolymerisation of poly(ferrocenylmethyl methacrylate)-*block*-poly(nonafluorohexyl methacrylate) (Figure 3.4), the nonafluorohexyl substituent on the methacrylate makes the pi bond in the methyl methacrylate even less nucleophilic, implying that the ferrocenylmethyl methacrylate block should be first synthesised before the nonafluorohexyl methacrylate block.[5]

Although less understood, the steric effect of organometallic moieties should alter the polymerisability of olefinic monomers. Indeed, steric strain decreases the enthalpy of polymerisation due to the interaction between substituents on the polymer. Organometallics such as ferrocene and cyclopentadienylmanganese tricarbonyl are bulky and could, therefore, decrease their monomer's polymerisability (discussed in Section 3.4).

3.3 Radical Chain Polymerisation

Initiators used in radical chain polymerisation are usually molecules with weak sigma bonds such as peroxides and disulfides. Molecules such as azo compounds that dissociate into more thermodynamically stable molecules also function as initiators. The initiation step involves homolytic cleavage of the initiator molecule into a pair of

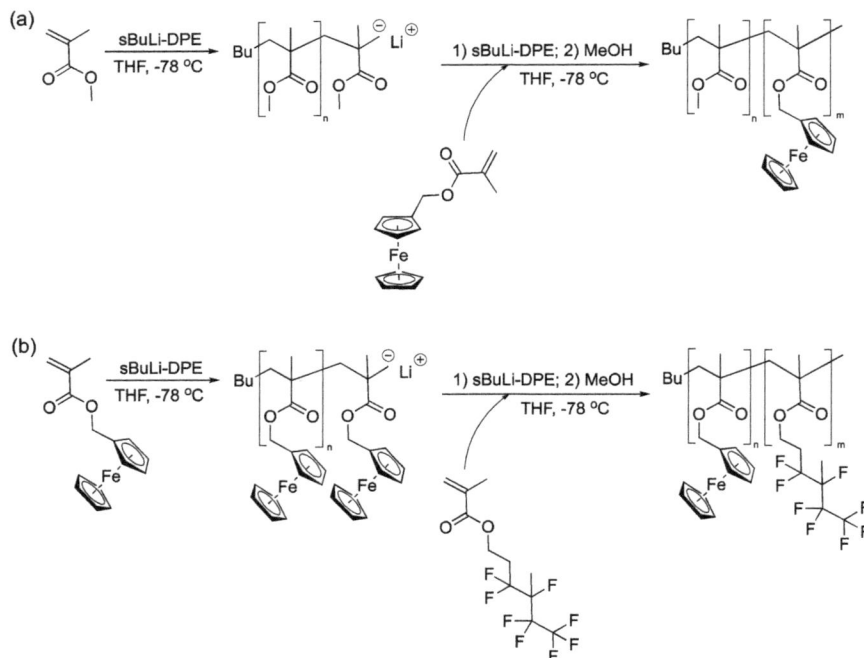

Figure 3.4 The effect of the organometallic moiety on the polymerisability of the pi-bond affects sequential copolymerisation. (a) In the synthesis of poly(ferrocenylmethyl methacrylate)-*block*-poly(methyl methacrylate) *via* sequential anionic chain polymerisation, the methyl methacrylate block is first synthesised before the ferrocenylmethyl block because the ferrocenylmethyl substituent lowers the nucleophilicity of the pi-bond. (b) In contrast, the ferrocenylmethyl methacrylate block is first synthesised before the nonafluorohexyl methacrylate block in the synthesis of poly(ferrocenylmethyl methacrylate)-*block*-poly(nonafluorohexyl methacrylate) because the nonafluorohexyl substituent makes the pi-bond even less nucleophilic.

primary radicals (Figure 3.5). Next, the primary radicals add to a monomer to form chain initiating radicals, which rapidly grows in the propagation step by successive addition of several monomers (Box 3.2). The propagation step consists of successive additions of monomers to the chain propagating radicals to produce the polymer (Box 3.2). Termination stops the growth of the polymer through intermolecular recombination of two propagating radicals or disproportionation reaction in which one propagating radical accepts a hydrogen atom from another propagating radical. Termination can also occur through the intermolecular reaction of the propagating radical with an added inhibitor or impurities such as oxygen and water (Box 3.2).

In some ferrocene-containing polymers, intramolecular electron transfer from the ferrocene to the propagating radical can terminate the polymerisation (Figure 3.3). Also, ferrocene can catalyse the decomposition of the initiator to inhibit the polymerisation. Nevertheless, several ferrocene-containing polymers are produced *via* radical chain polymerisation using azobisisobutyronitrile (AIBN) as the initiator species. AIBN-initiated radical chain homopolymerisation of vinylferrocene is distinct from that of organic vinyl monomers, such as styrene. With vinylferrocene the initial rate of polymerisation at low monomer conversion depends on the initial concentrations of monomer and initiator raised to the power of 1.1, whereas, with organic vinyl monomers, the rate depends

Initiation with azo compounds

$$R-N=N-R \xrightarrow[\text{Heat}]{N_2} 2\,R^{\bullet} \xrightarrow{R_1} \begin{array}{c} R \\ | \\ R_1 \end{array}^{\bullet}$$

Primary radicals   Chain initiating radical

Initiation with peroxides

$$R-O-O-R \xrightarrow{\text{Heat}} 2\,RO^{\bullet} \xrightarrow{R_1} \begin{array}{c} R-O \\ | \\ R_1 \end{array}^{\bullet}$$

Primary radicals   Chain initiating radical

Initiation with disulphides

$$R-S-S-R \xrightarrow{\text{Heat}} 2\,RS^{\bullet} \xrightarrow{R_1} \begin{array}{c} R-S \\ | \\ R_1 \end{array}^{\bullet}$$

Primary radicals   Chain initiating radical

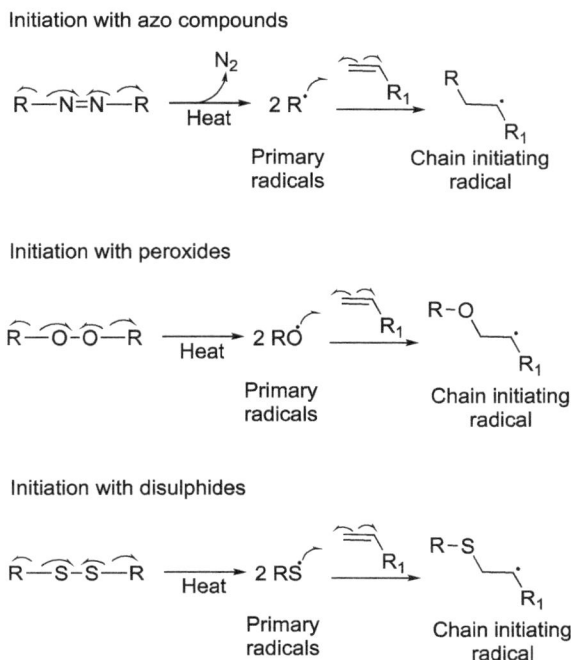

Figure 3.5 Some classes of compounds used as thermal initiators and their mechanism of forming primary and chain initiating radicals.

on the square root of the initiator concentration.[3] The polymerisation rate (R_p) of the vinylferrocene monomer (M) in the presence of an initiator (In) at 60 °C in benzene is defined by eqn (3.2). Nevertheless, the overall activation energy for the radical chain homopolymerisation of vinylferrocene at 60–80 °C in benzene is 139.7 kJ mol^{-1}, which falls within the range of most thermally initiated radical chain polymerisations.[3]

$$R_p = k'[M]^{1.1}[In]^{1.1} \tag{3.2}$$

3.3.1 Chemistry of Radical Initiators

It is crucial to consider the initiator's chemistry, which must be compatible with that of the organometallic monomer and initiating chain radicals to prevent adverse reactions that can prematurely terminate the polymerisation. Initiators are cleaved into primary radicals by thermal, photochemical, or redox stimuli. The low bond dissociation energies of the initiator, usually in the range of 100–170 kJ mol^{-1}, or the high stability of molecules such as nitrogen generated from the dissociation of the initiator, provide the driving force for thermal cleavage (Figure 3.5). Only a few molecules meet these criteria for thermal cleavage, and those with bond energies outside 100–170 kJ mol^{-1} either dissociate rapidly or require high temperatures to dissociate. Thermally initiated polymerisations use compounds having O–O bonds such as peroxides to generate primary radicals at temperatures ranging from 70 to 140 °C. However, some peroxides are decomposed by transition metals, including the iron atom in ferrocene. Another thermal initiator class is the family of azo compounds such as AIBN used in the radical polymerisation of several organometallic vinyl monomers, including vinylferrocene and vinylsilane. The bond dissociation energy of C–N in azo compounds is about 290 kJ mol^{-1}, which is higher than the typical 100–170 kJ mol^{-1} range for peroxide initiators. However, the extreme thermodynamic stability of the nitrogen generated during the dissociation of AIBN favours the formation of primary radicals (Figure 3.5). AIBN is one of the most used initiators and generates primary radicals at 50–70 °C. Another reason for the extensive use of peroxide- and azo-type initiators is their availability and stability under ambient conditions.

Ultraviolet or visible light can also cleave the initiator. Indeed, ultraviolet or visible light is one of the most convenient and greenest approaches to generate primary radicals for chain polymerisation. As a caveat, however, polymerisation involving the use of light, commonly termed photopolymerisation, should be avoided if the monomer or initiating chain radicals are light sensitive. Photoinitiated radical chain polymerisation can proceed *via* two mechanisms. In one mechanism, ultraviolet or visible light irradiation excites the photoinitiator and the excited state molecules decompose into two primary radicals to initiate polymerisation (Box 3.3). In the other mechanism, a photosensitiser

---

Box 3.3   Types of photoinitiators.

Type 1 photoinitiator: upon UV light absorption, these photoinitiators cleave to form two primary radicals, which can initiate the polymerisation. Examples include benzoin ethers, acetophenones, benzoyl oximes, and phosphine oxides.

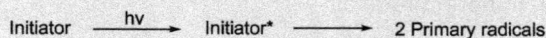

Example

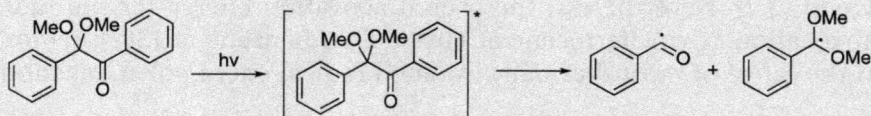

Type 2 photoinitiator: upon UV light absorption, these initiators abstract hydrogen from an electron donor (co-initiator) such as amines and alcohols, generating primary radicals. Only the radical from the donor initiates the polymerisation. Examples include benzophenones, quinones, and xanthones.

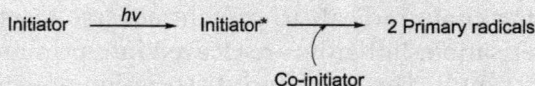

Example

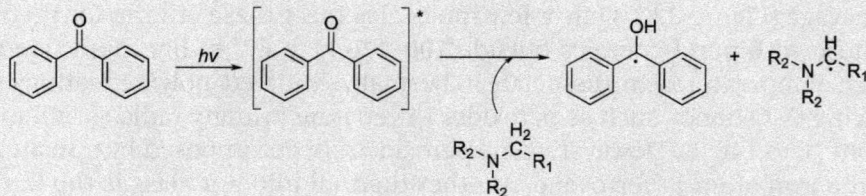

first absorbed the light energy to generate excited-state species that react with a co-initiator to generate primary radicals (Box 3.3). Photoinitiated radical chain polymerisation is well suited for heat sensitive monomers if the light energy is not converted to heat energy. The polymerisation can be spatiotemporally controlled by turning the light source on or off in specific time and space domains. The technique is applicable in the printing and photolithography industries, where photo-induced cross-linking of prepolymers or polymers enables ultrafast drying and patterning of polymeric inks and films (discussed in Chapter 6). For example, photo-induced cross-linking of poly-ferrocenylsilanes bearing pendant methacrylate groups was achieved using ultraviolet light irradiation of a photoinitiator, 2,2-dimethoxy-2-phenylacetophenone or phenylbis(2,4,6-trimethylbenzoyl)phosphine oxide.[6] In cross-linking reactions, the photoinitiator generates primary radicals upon absorption of ultraviolet light.

Other excitation methods such as the use of microwave, plasma, ionising, and high-intensity ultrasound radiations can dissociate the initiator to generate primary radicals. For example, sonication of gallium or eutectic gallium indium with a vinyl monomer results in radical chain polymerisation.[7] The sonication results in nanosized gallium particles, which have an increased surface area for interacting with the monomer. Gallium has the electronic configuration of $[Ar]3d^{10}4s^24p^1$, making it a source of an unpaired electron that acts as a primary radical to initiate chain polymerisation.

In redox-initiated radical chain polymerisation, an oxidising agent is mixed with a reducing agent to generate primary radicals. Oxidising agents used in these polymerisations typically have low bond dissociation energies and are usually molecules such as peroxides and disulfides with weak O–O and S–S sigma bonds. The low bond dissociation energies result in low activation energies (40–80 kJ mol^{-1}), which enable the polymerisation to proceed under mild conditions. An example of a redox initiator system is Fenton's reagent, a combination of hydrogen peroxide and an iron(II) salt. In this system, the iron(II) ion transfers one electron to the peroxide, dissociating the O–O sigma bond to generate one hydroxyl radical and one hydroxyl ion (Figure 3.6a). Fenton's reagent generates primary radicals to initiate reversible addition–fragmentation chain transfer (RAFT) polymerisation of vinyl monomers for the synthesis of well-defined polymers ($Đ < 1.10$).[8] Redox-initiated RAFT polymerisation is a promising method to synthesise polymers with controlled molecular weight under mild conditions such as room temperature and in the presence of air. The analogous reaction of a ferrous ion with halogens, disulfides, persulfates, or organic peroxides also generates primary radicals to polymerise vinyl monomers. Other reducing agents such as Ce^{4+}, Ag^+, Cr^{2+}, V^{2+},

(a) $Fe^{2+} + H_2O_2 \longrightarrow Fe^{3+} + HO^{\bullet} + OH^-$

$HO^{\bullet} + H_2C=CHR_1 \longrightarrow HOCH_2-\overset{\bullet}{C}HR$

(b) $Ce^{4+} + RCH_2CH_2OH \longrightarrow Ce^{3+} + RCH_2\overset{\bullet}{C}HOH + H^+$

$RCH_2\overset{\bullet}{C}HOH + H_2C=C(CH_3)COOCH_3 \longrightarrow RH_2C(HO)HC-H_2C-\overset{\bullet}{C}H(CH_2)COOCH_3$

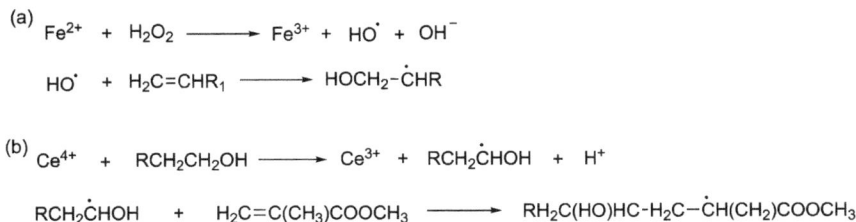

Figure 3.6 Redox-initiated radical chain polymerisation of vinyl monomers: (a) Fenton's reagent initiated and (b) Ce^{4+}/alcohol initiated radical chain polymerisation of vinyl monomers.

Ti^{2+}, Co^{2+}, and Cu^{2+} can generate radicals to initiate the polymerisation of vinyl monomer in aqueous or emulsion systems. For example, a Ce^{4+}/alcohol redox system (Figure 3.6b) initiates the radical chain polymerisation of the methyl methacrylate co-polymer over a broad temperature range (25–70 °C).[9]

In addition to the aforementioned inorganic redox initiating systems, an organic system comprising an organic peroxide, such as acyl peroxide, as the oxidising agent and an amine, such as 4-N,N-trimethylaniline, as the reducing agent can be used to generate primary radicals. One of these organic redox systems, for instance, the lauroyl peroxide/N,N-dimethylaminoethyl methacrylate redox system, initiates the radical chain copolymerisation of vinyl acetate with methacrylate monomers to form a branched copolymer.[10] Although the redox-initiated radical chain polymerisation proceeds under mild conditions, the use of toxic reagents such as peroxides and aromatic amines offset any advantage. Less toxic redox initiator systems such as a mixture of copper(II) acetylacetonate as the oxidising agent and 2-diphenylphosphinobenzoic acid as the reducing agent which initiates the polymerisation of methacrylate are attractive alternatives. The redox initiator system is yet to be used to polymerise organometallic monomers such as vinylferrocene, given that most of these monomers are redox-active. Nevertheless, the feasibility of a Cu^{2+}/amine redox system to initiate the radical chain polymerisation of methacrylate on titanium(IV) oxide nanoparticles to form inorganic/organic hybrid nanoparticles (Figure 3.7) highlights the potential of these systems in the synthesis of organometallic polymers.[11]

Electrolysis can also be used to generate primary radicals. An example is the electropolymerisation of nickel-, copper-, and zinc-salicylaldimine in dichloromethane or acetonitrile, where the metal atom oxidises, resulting in rapid intramolecular electron transfer.[12] The outcome is a reactive ligand-radical species that undergo oxidative polymerisation through radical coupling. It is important to note that oxidative polymerisation of metal salicylaldimine complexes is due to the ligand's redox chemistry; however, the metal's redox chemistry plays a vital role in the overall mechanism.

3.3.2 Reversible Deactivation Radical Polymerisation

A hallmark of radical polymerisation is the constant presence of bimolecular termination and chain transfer reactions. These reactions shorten the average lifetime of the initiating chain radicals to less than a second ($\sim 10^{-3}$ seconds). A strategy is to inhibit biomolecular termination and chain transfer reactions to prolong the lifetime of the radicals to hours or the entire duration of the reaction. In reversible deactivation radical polymerisation (RDRP), the lifetime is prolonged by keeping most initiating

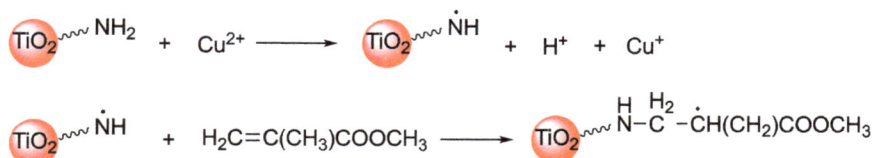

Figure 3.7 Copper/amine-initiation system generates a radical for the chain polymerisation of methyl methacrylate used to synthesise inorganic/organic hybrid nanoparticles.

chain radicals in a dormant state to lower the concentration of the reactive species and eventually renders bimolecular termination and chain transfer reactions negligible. Most literature referred to this polymerisation as "living" radical polymerisation. However, IUPAC discourages using the term "living" because it means a complete absence of termination and chain transfer reactions, a most unlikely phenomenon in a radical chain polymerisation. Another acceptable name to describe this polymerisation is "*controlled reversible-deactivation radical polymerisation*" or simply "*controlled polymerisation*" because the technique enables the precise synthesis of well-defined polymers of controlled molecular weights, dispersity, and architecture. This monograph will refer to this polymerisation as controlled polymerisation. The technique allows the synthesis of many organometallic and coordination polymers exemplified by the aliphatic salen polymer (Figure 3.8) with well-controlled molecular weight and molecular weight distribution ($Đ < 1.50$).[13]

Controlled polymerisation is exceptionally suited for the synthesis of block copolymers by sequential addition of different monomers. A typical example is the synthesis of block copolymers containing cobalt or iron (Figure 3.9).[14] In controlled polymerisation, the initiator dissociates into an equal concentration of a reactive radical and a stable radical. The reactive radical initiates the polymerisation, forming an initiating chain radical, commonly called the propagating chain radical, while the stable radical reacts with the propagating chain radicals to form dormant species (Figure 3.10). The dormant species and the propagating chain radicals are in equilibrium; however, the equilibrium favours the dormant species. The concentration of the dormant species is over six orders of magnitude higher than that of the propagating chain radicals. At any time, the majority of the propagating chain radicals are deactivated into a dormant state, with only a small fraction being active and growing into polymers. Activation and deactivation of the propagating chain radicals must be reversible and faster than the polymer growth rate to ensure that all chains are growing at the same rate. The overall results are that the low concentration of propagating chain radicals suppresses the termination step to a negligible level and ensures a narrow Poisson distribution and uniform dispersity. Controlled polymerisation proceeds *via*

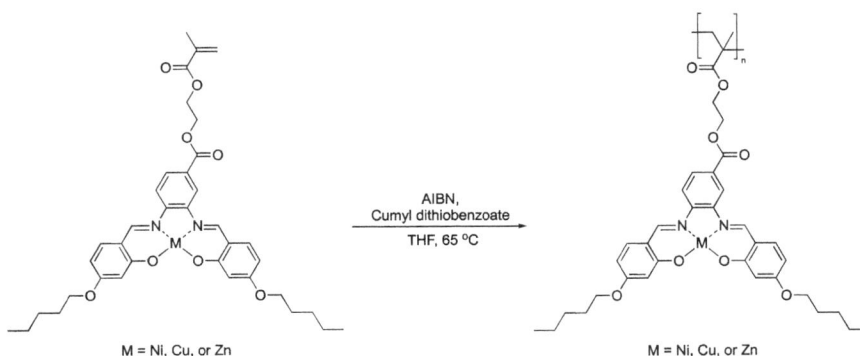

AIBN,
Cumyl dithiobenzoate
THF, 65 °C

M = Ni, Cu, or Zn   M = Ni, Cu, or Zn

Figure 3.8 Synthesis of coordination polymers ($Đ < 1.30$) *via* reversible addition–fragmentation chain transfer polymerisation, a type of reversible deactivation radical polymerisation. Cumyl dithiobenzoate (CDB) functions as a chain transfer agent and AIBN as an initiator.

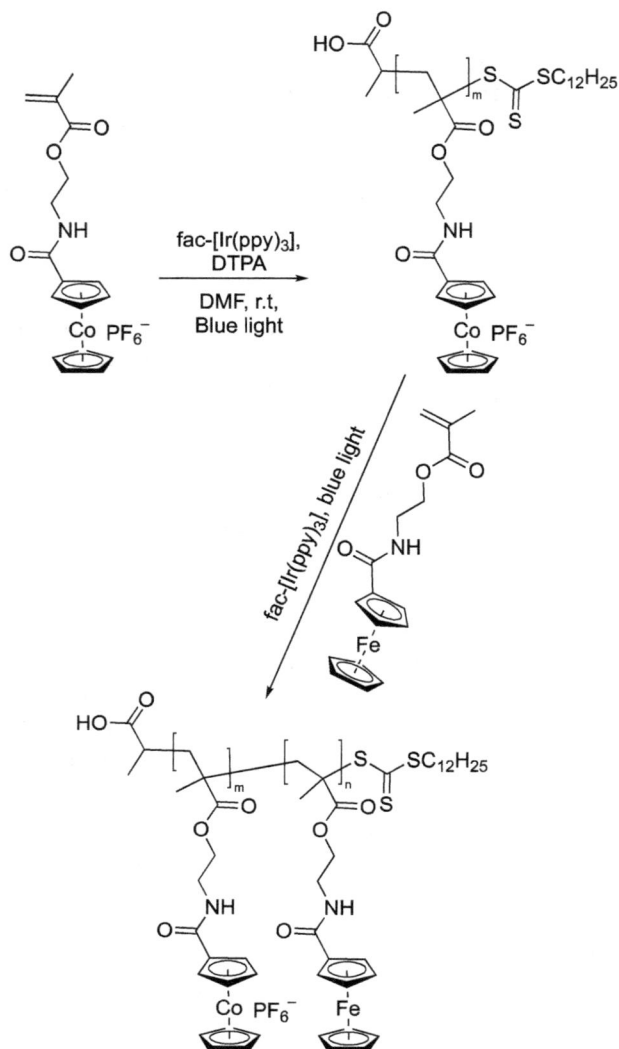

Figure 3.9 Sequential RAFT copolymerisation for the synthesis of cobalt- and iron-containing block copolymers. DTPA is (dodecylthiocarbonothioylthio)-2-methylpropionic acid and functions as a chain transfer agent. Fac-[Ir(ppy)$_3$] is tris[2-phenylpyridinato-C^2,N]iridium(III) and acts as a photoinitiator.

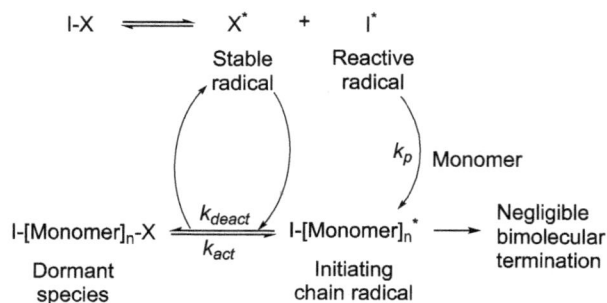

Figure 3.10 Mechanism of controlled polymerisation.

three mechanisms: atom-transfer, dissociation–combination, and degenerative chain transfer mechanisms (Box 3.4).

Box 3.4 History and mechanism of controlled radical polymerisation.

The industrial relevance of radical chain polymerisation fosters significant interest in developing a controlled version, now known as controlled polymerisation. The goal was to synthesise polymers with well-defined molecular weights, dispersity, and architectures. Successes recorded with controlled ionic chain polymerisation promoted the interest in developing controlled radical polymerisation. However, the prevalence of bimolecular termination reactions under typical radical polymerisation conditions makes precise control challenging. In 1982, however, the feasibility of a controlled radical polymerisation was demonstrated by Otsu and Yoshida with styrene and methyl methacrylate monomers, although there was no precise control of molecular weight.[15] Between 1993 and 1994, reports of organocobalt complex and 2,2,6,6-tetramethyl-1-piperidinyl-1-oxy (TEMPO) mediated radical polymerisations having RDRP characteristics as demonstrated with controlled molecular weight and dispersity emerged.[16] A 1999 report shows TEMPO-mediated radical polymerisations of vinylferrocene without controlling molecular weight and dispersity.[17] The lack of control is due to ferrocene acting as a chain transfer agent and the intramolecular electron transfer from the ferrocenyl iron atom to the propagating chain radical,[3] processes that prematurely terminate the polymerisation. Previously, in the 1970s, Pittman and coworkers found that inserting an alkyl spacer between the ferrocene and the polymerisable vinyl group mitigates the intramolecular electron transfer process.[18] The Pittman strategy now guides the design of ferrocene-containing monomers such as ferrocenyl methyl acrylate for controlled polymerisation.

   Mechanistically, three types of controlled polymerisation currently exist, and they differ by their mode of reversible activation of the dormant species.

1. *Atom transfer mechanism.* The dormant species (I-[Monomer]$_n$-X) is activated through a transition metal complex-catalysed reversible redox reaction with an organic halide. Typical transition metal complexes include those of Cu or Ru, while the organic halides are bromides and chlorides. For this mechanism to yield polymers of uniform dispersity ($Đ < 1.5$), the initiation step must be fast and quantitative to ensure the simultaneous growth of propagating species. The reversible deactivation step must also be fast to maintain low concentrations of radicals and minimise biomolecular termination. Together, these requirements ensure that propagating chains grow at the same rate and have the same lifetime.

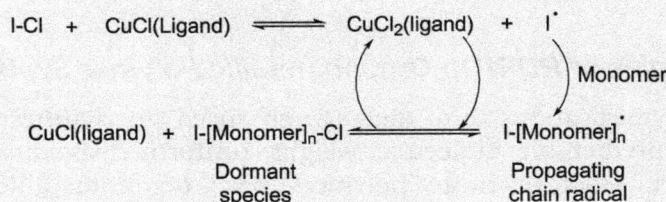

I-Cl + CuCl(Ligand) ⇌ CuCl₂(ligand) + I˙

Monomer

CuCl(ligand) + I-[Monomer]$_n$-Cl ⇌ I-[Monomer]$_n$˙

Dormant species   Propagating chain radical

2. *Dissociation–combination mechanism.* The dormant species (I-[Monomer]$_n$-ONR$_1$R$_2$) dissociate thermally or photochemically into a propagating radical (I[Monomer]$_n$) and a persistent radical (ONR$_1$R$_2$). A mixture of a conventional radical initiator such as AIBN and a nitroxide radical initiates the polymerisation. A unimolecular initiator such as an alkoxyamine initiator ensures better control over molecular weight and dispersity than a mixture of a radical initiator and a nitroxide radical. The persistent radical is stable enough to be unreactive, except towards the propagating radical. Typical persistent radicals include nitroxide and dithio-carbamate. A controlled polymerisation mediated by nitroxide is called nitroxide-mediated polymerisation (NMP). Compared to atom transfer and degenerative chain transfer mechanisms, the dissociation–combination mechanism is less versatile with respect to the number of monomers that undergo polymerisation, but the polymers are free from transition metals and odorous contaminants.

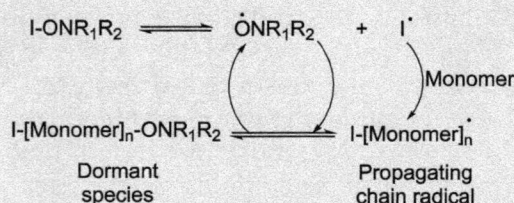

3. *Degenerative chain transfer mechanism.* The dormant species (I-[Monomer]$_n$-X) reacts with another propagating radical (R$_1$[Monomer]$_m$) to form the active species (I-[Monomer]$_n$) and a new dormant species (R$_1$[Monomer]$_m$-X). Two types of degenerative chain transfer mechanism exist. In one type, X is a simple group such as an iodine atom that is exchanged between radicals, forming a kinetically important intermediate as exemplified by iodine-mediated living radical poly-merisation. In the other type, X has a pi bond that reacts with R$_1$[Monomer]$_m$, forming an intermediate radical that fragments into I-[Monomer]$_n$ and R$_1$[Monomer]$_m$-X. This type is known as reversible addition–fragmentation chain transfer (RAFT) polymerisation. Dithioesters are typical RAFT agents, which, combined with a conventional radical initiator such as AIBN, are widely used in RAFT polymerisation. The polymerisation accommodates more monomers than the other living polymerisation mechanisms.

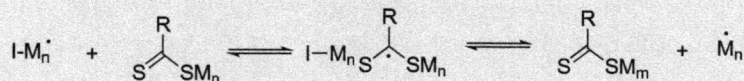

3.3.2.1   Examples of RDRP in Organometallic Polymer Synthesis

Controlled polymerisations are increasingly used to synthesise organometallic polymers with predictable molecular weight, uniform dispersity, and well-defined macrostructures. Most of these polymers bear organometallic motifs, such as metallocenes, as a side-chain group. The versatility of atom-transfer radical polymerisation (ATRP) and reversible addition–fragmentation chain transfer (RAFT) polymerisation (discussed in Box 3.4) makes them particularly useful in synthesising well-defined styrene- and methacrylate-based organometallic polymers. In contrast, nitroxide-mediated polymerisation (NMP) (discussed in Box 3.4) is less versatile, unable to polymerise methacrylate monomers into well-defined polymers due to the chain termination mediated by the nitroxide radical abstracting a β-hydrogen from propagating chain radical. Still, NMP, specifically a TEMPO-mediated polymerisation of vinylferrocene, is successful and yields poly(vinylferrocene) characterised by M_w and Đ ranging from 2100 to 4800 g mol^{-1} and 1.24 to 1.84, respectively.[17] Although the TEMPO-mediated polymerisation of vinylferrocene is successful, the mechanism deviates from a controlled polymerisation. For instance, the gel permeation chromatography (discussed in Chapter 4) estimated molecular weights are lower than the theoretical molecular weights, and the molecular weight distributions are broad and increase with the conversion and concentration of vinylferrocene. TEMPO-mediated polymerisation allows the block copolymerisation of vinylferrocene with styrene (Figure 3.11), but the vinylferrocene block is short due to its low conversion. The low conversion of vinylferrocene during TEMPO-mediated polymerisation is attributed to the stability of the TEMPO–vinylferrocene bond, which lowers the concentration of the propagating chain radicals and prematurely terminates the polymerisation of vinylferrocene. Another explanation for the low conversion of vinylferrocene during radical polymerisation is the intramolecular charge transfer from the ferrocene's iron to the propagating radical centre (see Section 3.3 and Figure 3.3). A strategy is to design a monomer in which the propagating radical centre is spatially distant and electronically isolated from the iron centre (Box 3.4). This strategy enables nitroxide-mediated polymerisation of a styrene monomer conjugated to an iridium(III) complex to form iridium-containing polymers characterised by $M_n = 30\,000$–$150\,000$ and Đ < 1.5 (Figure 3.12).[19]

Unlike NMP, ATRP and RAFT polymerisation tolerate different monomers and functional groups and exhibit better control over polymerisation kinetics. These

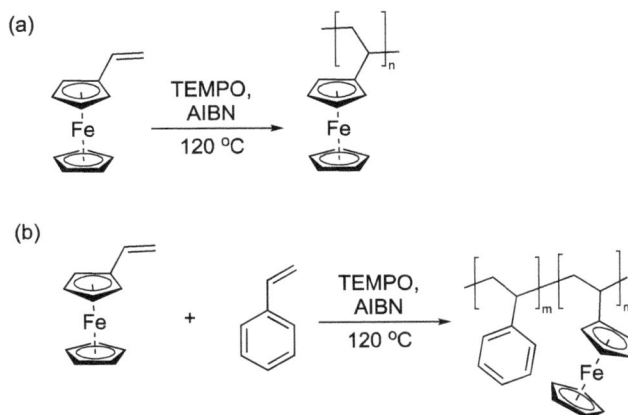

Figure 3.11 TEMPO-mediated polymerisation. (a) Homopolymerisation of vinylferrocene and (b) copolymerisation of vinylferrocene with styrene.

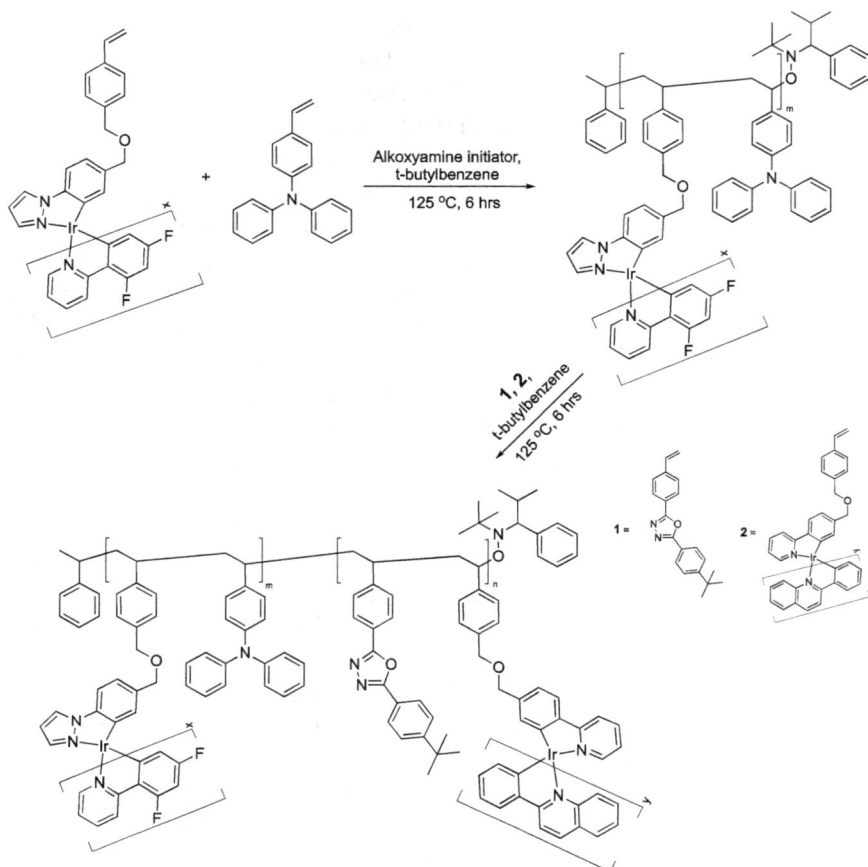

Figure 3.12 Nitroxide-mediated polymerisation of styrene monomer conjugated to an iridium complex.

features informed the prolific use of ATRP and RAFT polymerisation in the controlled radical polymerisation of organometallic monomers. Another reason for the popularity of ATRP and RAFT polymerisation is their versatility in terms of polymerisation conditions. These polymerisations proceed in bulk or in various organic solvents, including toluene, THF, DMF and 1,4-dioxane, and occur over a wide range of temperatures. For instance, RAFT polymerisation of a nickel-containing salen complex (Figure 3.8) proceeds in THF at 65 °C,[13] whereas that of a ruthenocene-containing methacrylate monomer (Figure 3.13) occurs in 1,4-dioxane at 90 °C or DMF at 60 °C.[20] In both cases, the kinetics are characteristics of controlled radical polymerisation as evidenced by the narrow molecular weight distribution ($Đ < 1.16$), the linear relationship between $\ln([M]_0/[M])$ and reaction time, and the linear relationship between molecular weight and monomer conversion (Figure 3.14). The ruthenocene-containing methacrylate monomer also undergoes ATRP to produce homo and block copolymers (Figures 3.13 and 3.14).[20] For the homopolymerisation of the ruthenocene-containing monomer, ATRP is a more efficient technique than RAFT polymerisation, providing better control over polymerisation even at high monomer conversion. For RAFT

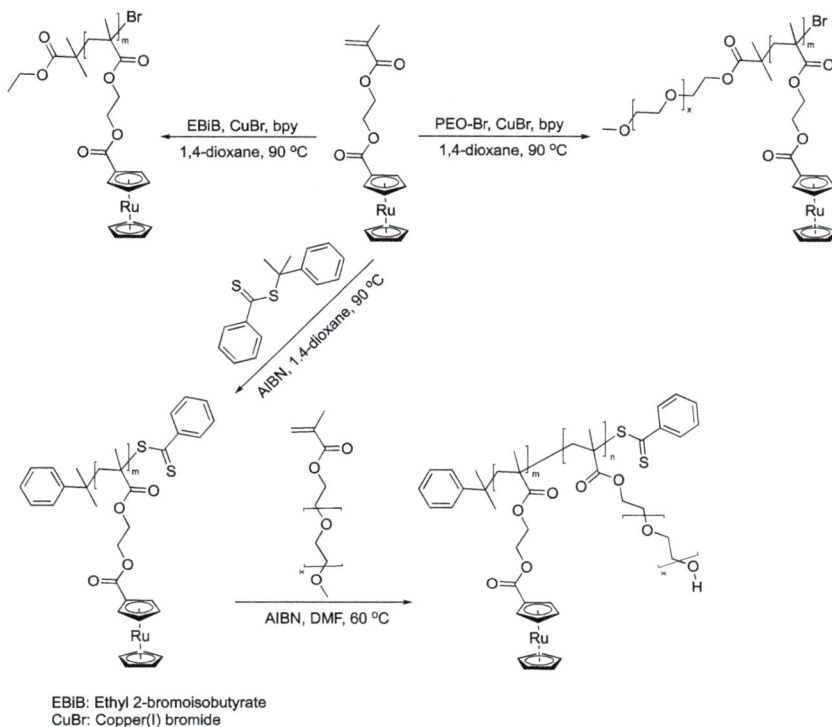

Figure 3.13 Synthesis of a ruthenocene-containing methacrylate monomer *via* ATRP and RAFT polymerisation.

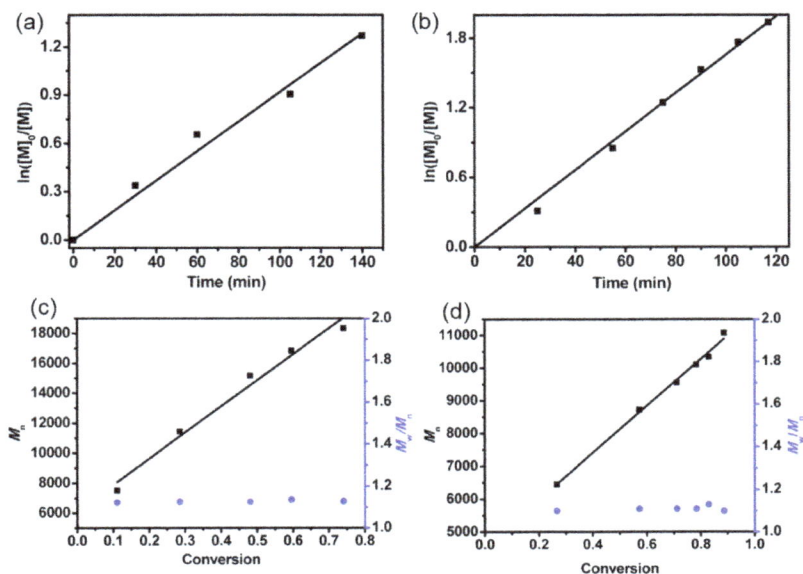

Figure 3.14 RAFT polymerisation (a and c) and ATRP (b and d) of ruthenocene methacrylate (Figure 3.13) follow controlled radical polymerisation kinetics as evidenced by the linear relationship between $\ln([M]_0/[M])$ and reaction time (a and b), the linear relationship between molecular weight and monomer conversion and the low dispersity ($Đ = M_w/M_n$) (c and d). Reproduced from ref. 20 with permission from American Chemical Society, Copyright 2013.

polymerisation, the kinetics deviates from controlled radical polymerisation as the monomer conversion exceeds 72%. Loss of control at very high monomer conversion occurs in all controlled radical polymerisations.

ATRP and RAFT polymerisation of other metallocene-containing (meth)acrylate monomers also follow controlled radical polymerisation kinetics. ATRP is the most versatile among controlled radical polymerisations considering its ability to proceed under different conditions and tolerate various monomers and initiators. Different initiator systems initiate the ATRP of ferrocene-containing (meth)acrylate to produce polymers of different macrostructures. For example, an ethyl α-bromoisobutyrate initiator produces a homopolymer, whereas a poly(ethylene glycol) macroinitiator gives a block copolymer (Box 3.5). On the other hand, a graphene oxide macro-initiator yields a polymer-graphene oxide nanocomposite (Figure 3.15).[21]

Box 3.5   Worked example 3.1.

Question

   Present an ATRP strategy to synthesise a poly(ferrocenylmethyl methacrylate)–fullerene nanocomposite.

Answer

   First, synthesise a brominated fullerene from fullerol. Next, use the brominated fullerol in the ATRP of ferrocenylmethyl methacrylate.

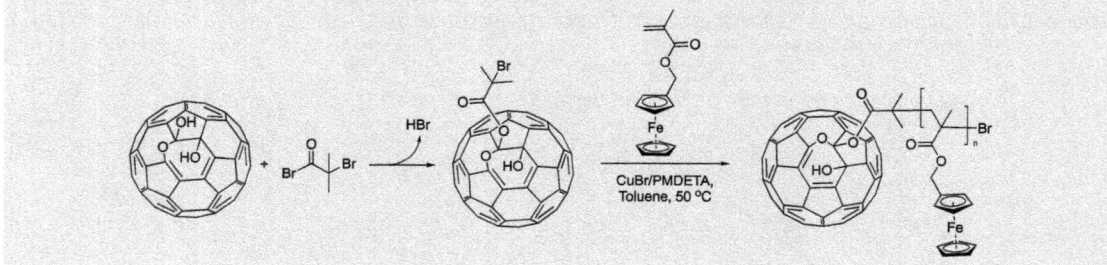

3.4   Ionic Chain Polymerisation

Ionic chain polymerisations are highly selective reactions, requiring monomers with appropriate electronic effects and solvents with suitable polarity. Both requirements should be satisfied to stabilise the propagating ionic species and ultimately prolong the lifetime of initiating chain ionic species. Ionic chain polymerisations are classified into cationic and anionic chain polymerisations. For cationic chain polymerisation, a prerequisite is that the monomers contain an electron-donating substituent that stabilises the initiating cationic species. On the other hand, monomers in anionic chain polymerisation require an electron-withdrawing substituent to stabilise the propagating anionic species. Ideally, high polarity solvents, such as water and methanol, that solvate the ionic propagating species should be preferred because they stabilise the propagating ionic species. However, these solvents decompose most

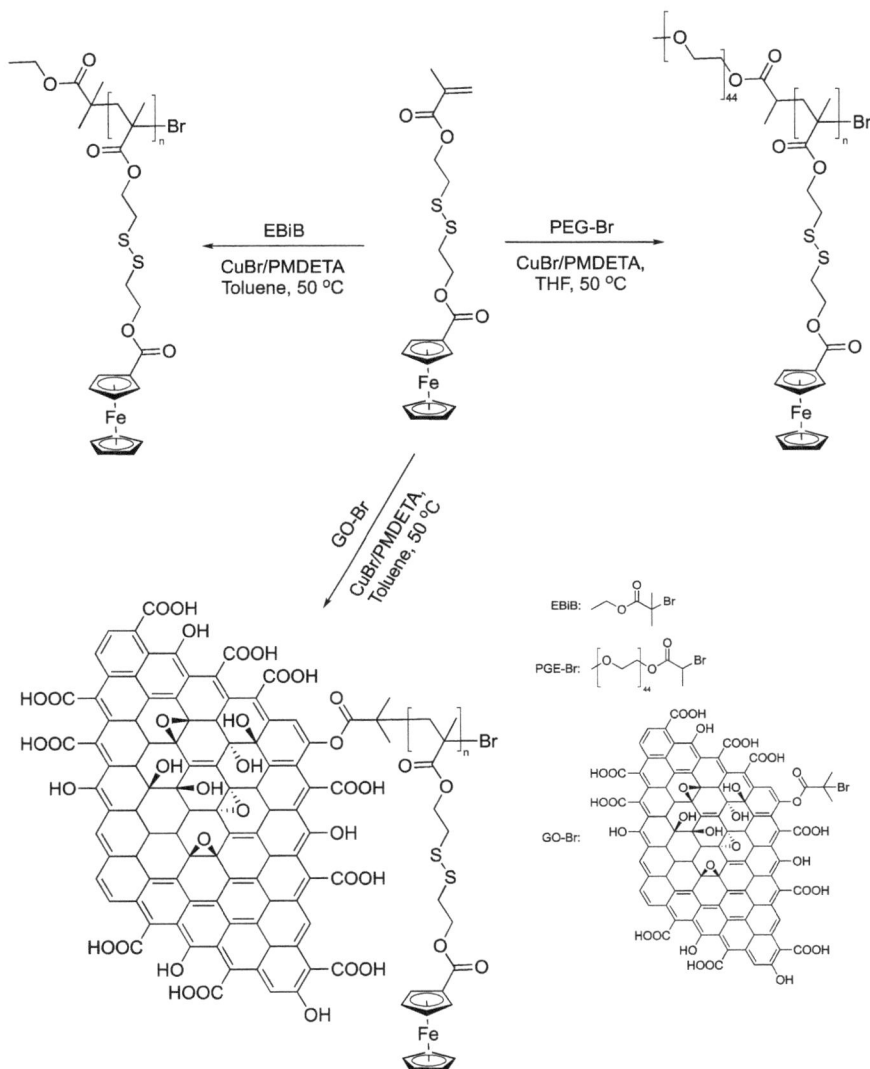

Figure 3.15 Synthesis of a ferrocene-containing homopolymer, block copolymer, and graphene oxide nanocomposite *via* ATRP.

initiators or complex the initiator, precluding the initiation step. Therefore, mildly polar solvents, such as pentane, tetrahydrofuran, and ethylene dichloride, are used in ionic polymerisation. The polymerisation also requires moderate to low temperatures to suppress chain termination reactions that annihilate the initiating chain ionic species. In a mildly polar solvent, the initiating ionic species can exist as completely covalent molecules, tight ion pairs, loose ion pairs, and free ions with the ion pair and free ion in equilibrium depending on the solvent polarity and the size of the counterion (Box 3.6). Termination of ionic polymerisation occurs by reaction of the propagating ionic species with the counterion, solvent, or other impurities present in the reaction vessel.

Box 3.6 Worked example 3.2.

Question

   Discuss the effect of solvent polarity and counterion on forming a tight and loose ion pair and its implications for cationic and anionic polymerisations.

Answer

   Polar solvents can solvate ions, so increased solvent polarity favours loose ion pairs – initiating ionic species are partially solvated, separated from the counterion. Conversely, decreased polarity favours tight ion pairs – initiating ionic species are not solvated, so bound to the counterion. A large counterion favours loose ion pair formation since the low charge density weakens electrostatic interaction with the initiating ionic species. On the other hand, a small counterion has a high charge density strengthening the electrostatic interaction with the initiating ionic species, resulting in tight ion pair formation. In cationic polymerisation, the counterions are typically large, for example, a perchlorate ion, so the ion pair tends to be loose. In anionic polymerisation, the counterions are small, such as lithium ions; therefore, the ion pair tends to be tight.

Electron-donating organometallic vinyl monomers, such as vinylferrocene and vinylcymantrene, undergo cationic chain polymerisation because they readily form stable initiating cationic species. These organometallic monomers are far more reactive than organic vinyl monomers such as styrene during cationic polymerisation.[22] The cyclopentadienyl ring in these organometallic monomers is more electron-rich than a phenyl ring due to pi-back donation of electrons from the d-orbitals of the transition metal. Consequently, in the cationic copolymerisation of styrene and vinylferrocene, the latter monomer is more readily incorporated into the copolymer than the former. Styrene is incorporated only when its monomer feed exceeds 90% of the total monomer feed.[22] In addition to the electronic effect, the steric bulk of the organometallic moiety also influences the rate of polymerisation. For instance, in the cationic copolymerisation of vinylferrocene and vinyl isobutyl ether, the former monomer is less readily incorporated into the copolymer than the latter.[22] In this case, although the alkoxy group is strongly electron-releasing, the steric bulk resulting from coupling many ferrocenyl groups along the polymer backbone decreases the reactivity of vinylferrocene relative to vinyl isobutyl ether.

Generally, Brønsted–Lowry acids and Lewis acids initiate the cationic polymerisation of monomers with electron-donating substituents. For Brønsted–Lowry acids, the acid must be sufficiently strong to protonate the monomer, but its conjugate base should not be excessively nucleophilic to terminate the polymerisation *via* recombination with the propagating cationic species. This requirement limits the use of strong Brønsted–Lowry acids as an initiator due to the strong nucleophilicity of their conjugate bases. Some Brønsted–Lowry acids such as hydrogen iodide and sulfuric acid with weak nucleophilic conjugate bases initiate cationic polymerisation, but the degree of polymerisation is generally low. On the other hand, Lewis acids efficiently initiate cationic polymerisation, even at low temperature, to form high molecular weight polymers. To be specific, the

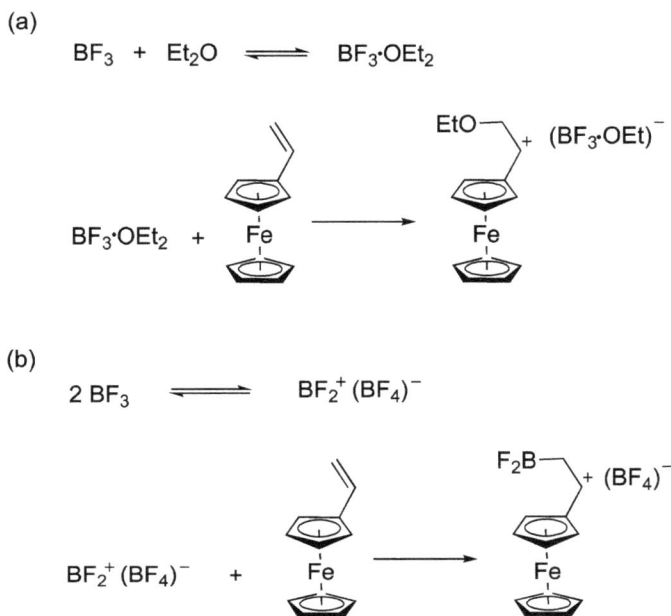

Figure 3.16 Lewis acid initiation of cationic polymerisation of the organometallic polymer exemplified by vinylferrocene. (a) Lewis acid acts as a co-initiator with the diethyl ether initiator. (b) Lewis acid self-ionises, acting as the initiator and co-initiator.

Lewis acid functions as a co-initiator, synergising with the initiator, which is usually a proton donor or a carbocation donor. These donors are called initiators because they furnish the proton or cation that adds to the monomer, initiating the polymerisation (Figure 3.16). In certain conditions, under extremely dried reaction procedures that lack proton donors such as water, the Lewis acid self-ionises, acting as the initiator and co-initiator. However, these conditions are rare since, in many polymerisations, the water content is enough to reduce self-ionisation to a negligible degree. In any case, the initiator reacts with the co-initiator to form an initiating system whose activity depends on the ability of the system to add a proton or cation to the monomer. The initiator, therefore, plays a pivotal role in the overall success of polymerisation. For instance, in the cationic polymerisation of vinylferrocene, the monomer conversion and polymer molecular weight depend on the initiating system (Table 3.1).[22]

Theoretically, electron-rich organometallic vinyl monomers should resist anionic chain polymerisation because the electron-donating substituents destabilise an initiating anionic species. However, these monomers undergo anionic polymerisation because electron delocalisation from the initiating anionic species to the d-orbital of the transition metal makes the organometallic moiety act like an electron-withdrawing group. Indeed, anionic chain polymerisation of vinylferrocene is feasible using butyllithium as the anionic initiator and tetrahydrofuran as the solvent (Figure 3.17).[23] The success of organolithium-initiated anionic chain polymerisation is solvent-dependent. In nonpolar solvents such as toluene and benzene, lithium species associate, forming an unreactive species that precludes the initiation, slowing the polymerisation rate. The association of lithium species could be inhibited by selecting an appropriate solvent. The association disaggregates when polymerisation is carried

Table 3.1   Effect of the initiating system on cationic polymerisation of vinylferrocene in dichloromethane at 0 °C.

Initiating system	Conversion (%)	Mol. wt $(g\,mol^{-1})$
BF_3OEt_2	35	1500
$TiCl_4$	63	1100
$SnCl_4$	6	
$AlCl_3$	82	1700
$EtAlCl_2-^tBuCl$	61	1100
Et_2AlCl-^tBuCl	28	3500

Box 3.7   Worked Example 3.3.

Question

The cationic chain polymerisation of vinyl ruthenocene was carried out in dichloromethane consisting of BF_3 and a trace amount $(10^{-3}\,M)$ of water impurity. Will the BF_3 co-initiate the polymerisation with water or self-ionise to act as a co-initiator and an initiator?

Answer

A trace amount $(10^{-3}\,M)$ of water is sufficient to initiate the polymerisation, acting as the proton donor to the double bond of the polymerisable vinyl group. The BF_3 will co-initiate the polymerisation with water. The initiation proceeds as follows:

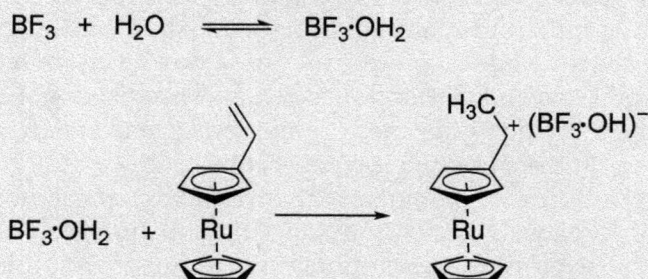

out in a polar solvent such as tetrahydrofuran or by adding a Lewis base such as ethers and amines or lithium chloride. For example, the organolithium-initiated anionic chain polymerisation of vinylferrocene in dioxane becomes possible only if a macrocyclic crown ether such as [12]-crown-6 is added to the reaction system.[23] Also, the dispersity of poly(ferrocenylmethyl methacrylate) obtained *via* controlled anionic polymerisation remarkably is improved from 1.30 to 1.03 by adding lithium chloride to the polymerisation solvent.[24]

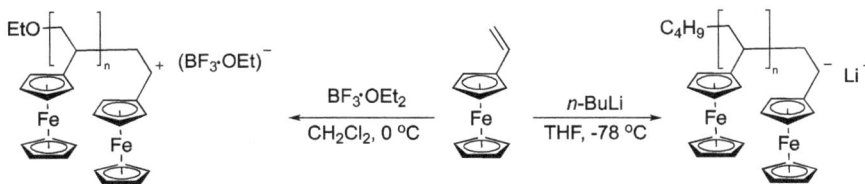

Figure 3.17 Vinylferrocene undergoes cationic and anionic polymerisations initiated by BF$_3$OEt$_2$ and n-BuLi, respectively.

Controlled anionic chain polymerisation is easily achievable in an extremely purified reaction system, devoid of impurities that terminate the propagating species. This contrasts with cationic chain polymerisation with built-in chain breaking reactions such as β-proton transfer to the monomer or reaction with the counterion that terminates the polymerisation. Nevertheless, controlled cationic chain polymerisation is feasible by judicious selection of the initiator, co-initiator, and reaction temperature. By and large, most organometallic vinyl monomers undergo cationic and anionic polymerisations (Figure 3.17).

3.5 Ring-opening Polymerisation

Ring-opening polymerisation (ROP) is a type of chain polymerisation where cyclic monomers open, forming a linear polymer. Various cyclic monomers, ranging from entirely inorganic molecules such as sulfur through organic–inorganic hybrids such as phosphonates to organometallics such as bridged metallocenes, undergo ring-opening polymerisation (Figure 3.18). ROP can proceed by a radical, cationic, or anionic mechanism depending on the monomer and initiator. Whether a given monomer will undergo ROP depends on the thermodynamics – the relative stability of the cyclic monomer and the resulting polymer – and the kinetics – the susceptibility of the monomer to react and open under the experimental conditions. The thermodynamic stability of cyclic monomers depends on ring strain that includes angle strain, torsional strain, and transannular strain (Box 3.8). As ring strain increases, thermodynamic stability decreases, improving the enthalpic driving force for ROP of cyclic monomers.

Although ROP can be thermodynamically feasible for a particular monomer, actual polymerisation depends on kinetic feasibility, which requires a monomer that is susceptible to an initiator's attack. A heteroatom or an unsaturated bond in a ring provides a reactive site for the nucleophilic, electrophilic, or radical attack needed to initiate and propagate the ROP (Figure 3.18). Polymerisability of cyclic monomers, therefore, requires both thermodynamic and kinetic feasibilities. For example, in a series of [2]ferrocenophane monomers (Figure 3.19) with unsymmetrical C–ER$_x$ bridges, changing E, a second-row heteroatom, and R$_x$, an organic group, alters the angle strain to control the polymerisability of the monomer. A heteroatom in these monomers does not guarantee ROP when there is insufficient ring strain (Figure 3.19). In [n]metallocenophanes, ring strain positively correlates with the angle (α) between the planes of the cyclopentadienyl rings (Box 3.8). Increasing the ring strain by replacing SiMe ($\alpha = 11.8(1)°$) with CH$_2$ ($\alpha = 21.6(5)°$), PPh ($\alpha = 14.9(3)°$), or S ($\alpha = 18.5(1)°$) yields [2]ferrocenophanes that undergo ROP (Figure 3.19).[25]

The presence of heteroatoms, highly strained structures or both does not assure successful ROP of organometallic monomers. The experimental conditions of the ROP

(a)

(b)

(c)

(d)

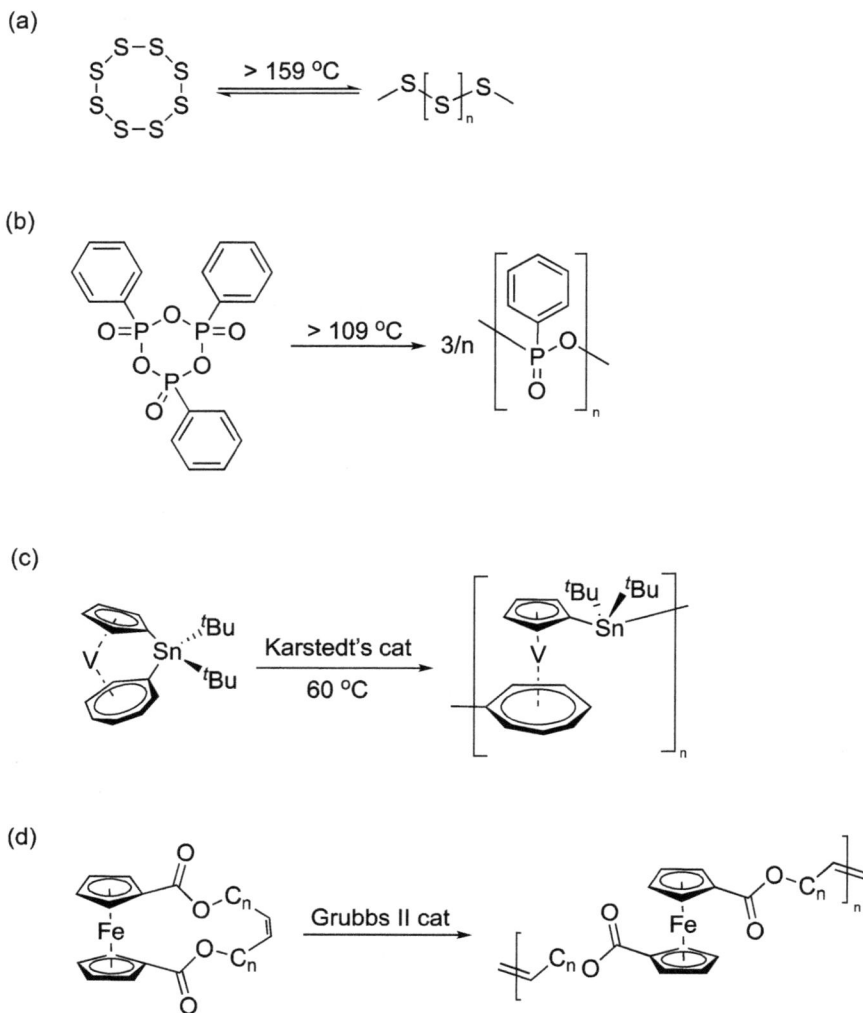

Figure 3.18   Ring-opening polymerisation of orthorhombic sulfur, cyclic phosphonate, and tin-bridged vanadoarenophanes and ring-opening metathesis polymerisation of cyclic ferrocenyl olefin.

Box 3.8   Ring strain and polymerisability of cyclic monomers.

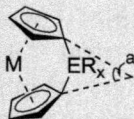

   In [n]metallocenophanes, the ring strain positively correlates with the angle (α) between the planes of the cyclopentadienyl rings. Generally, strain in cyclic structures can result from angle strain, torsional strain, or transannular strain. Angle strain (also known as Baeyer strain) results when the bond angle deviates from the ideal bond angle expected to achieve the maximum bond strength in each conformation. Angle strain plays a significant role in the thermodynamic stability of cyclic molecules that lack the flexibility of acyclic structures. In cyclopropane, for example, the tetra-coordinate carbon atoms should ideally have a tetrahedral bond angle of 109.5° but becomes distorted to 60° and highly strained as the structure assumes a rigid planar conformation to achieve maximum bond strength. In cyclic structures consisting of less than five-membered atoms, bond angles are significantly distorted from the ideal, resulting in a highly strained bond angle that decreases the thermodynamic stability and favours ROP. Cyclic structures of five or more member atoms exist in flexible, puckered forms that are stable and virtually free of angle strain. In these structures, ring strain is due to torsional strain, also known as Pitzer strain, that arises when adjacent atoms are in eclipsed conformation or transannular strain (also known as Prelog strain) that results from unfavourable interactions between crowded ring sub-stituents in the interior of the ring. Compared with cyclic structures of six-membered atoms, those with five- and seven-membered atoms have a torsional strain, while those with eight or more members have transannular strain. Ring strain in cyclic structures usually results from a combination of angle strain, torsional strain and transannular strain and generally follows the order of $3 > 4 > 5-8 > 6$. Increased ring strain de-stabilises the cyclic molecules to decrease the thermodynamic stability in the following order: $3 < 4 < 5-8 < 6$. This order is generally observed for various families of cyclic compounds including those containing heteroatoms. Polymerisability is, therefore, thermodynamically favoured and feasible in highly strained cyclic monomers.

process must be compatible with the chemistry of the monomer and the resulting polymer. Most organometallic compounds are highly labile and therefore likely to engage in various reactions, generating undesirable chemical species. For example, attempts to initiate thermal or anionic ROP in the highly strained [1]ferrocenophane ($\alpha = 24.5-28.4°$) containing a heteroatom, nickel, palladium or platinum, were un-successful.[26] These [1]ferrocenophanes resist ROP, dissociating into small molecule compounds (Figure 3.20). Under the experimental conditions of ROP including high temperature (200 °C), or the presence of phenyllithium in tetrahydrofuran, nickela- and pallada-[1]ferrocenophanes dissociate, while their platinum congener remains inert, phenomena that concord with their metal–carbon bond strength.

The ring-opening polymerisability of cyclic organometallic monomers depends on a complex interplay of the presence of a heteroatom, the degree of ring strain, and the reactivity of the metal–carbon bond. This is evidenced by the non-polymerisable behaviour of the highly strained group 10 metal-containing [1]ferrocenophanes (Figure 3.20), which contrasts with that of other less strained [1]metallocenophanes that readily undergo ROP to produce high molecular weight polymers (Figure 3.21). The less strained sila[1]ferrocenophane ($\alpha = 19.2°$) undergoes thermal ROP at 130 °C to produce a highly stable organometallic polymer characterised by a M_n of 340 000 Da (Figure 3.21).[27] Polymerisability also depends on the substituents or ligand on the heteroatom bridge in a [n]metallocenophane with a bulkier phenyl derivative requiring higher temperatures for ROP compared with its methyl congener (Figure 3.21).[27] Also, the role of the heteroatom bridge ligand in the ROP of the [n]metallocenophane is demonstrated by the non-polymerisability of the galla[1]ferrocenophane ($\alpha = 15.8°$), which has a similar

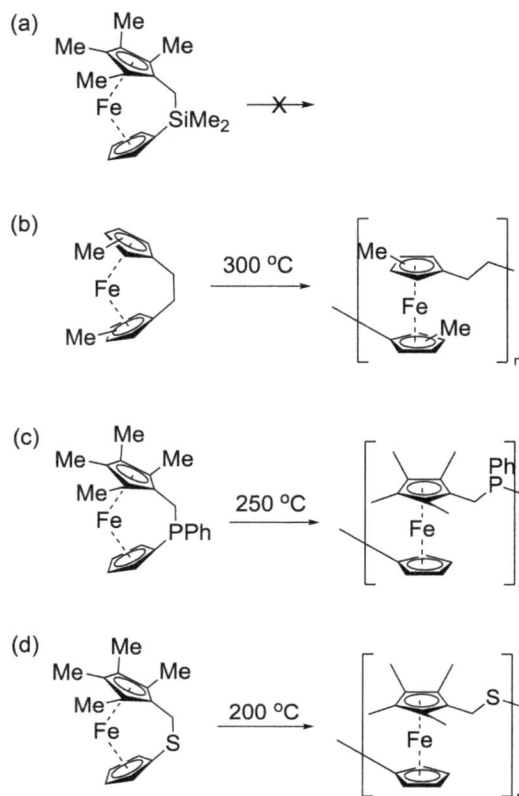

Figure 3.19   The presence of a heteroatom does not guarantee the ROP of [2]ferrocenophanes. All the [2]ferrocenophanes (a–d) have heteroatoms in their cyclic structure, yet only (b–d) undergo ring-opening polymerisation.

degree of ring strain and heteroatoms to the polymerisable galla[1]ferrocenophane ($\alpha = 15.7°$) (Figure 3.22).[28] In a non-polymerisable galla[1]ferrocenophane (Figure 3.22a), replacing an electron-donating dimethylamino group with an electron-withdrawing pyridyl group results in ROP (Figure 3.22b). This trend indicates that the electronic effects of the substituent on the bridging atom contribute to the kinetic feasibility of ROP of [n]metallocenophanes.

   The thermodynamic feasibility of ROP depends on ring strain, which results from an angle, torsional, or transannular strain (Box 3.7). These strains contribute to the enthalpic component (ΔH_{ROP}) of the free energy of the polymerisation reaction. As the ring size increases, the angle strain decreases its contribution to ΔH_{ROP}, while torsional and transannular strains increase their contributions. In cyclic structures such as [n]metallocenophanes, the angle strain decreases with increasing ring size. The low angle strain found in large rings such as tetracarba[4]nickelocenophane does not preclude their ROP because torsional strain becomes relevant in large rings and contributes more ΔH_{ROP}. Also, the entropic component of the free energy of ROP becomes favourable for large cyclic structures. Indeed, tetracarba[4]nickelocenophane, which possesses an insignificant angle strain ($\alpha = 1.0°$), undergoes ROP due to torsional strain.[29] The role of torsional strain in ROP is further substantiated by the inability of other [4]nickelocenophanes which have

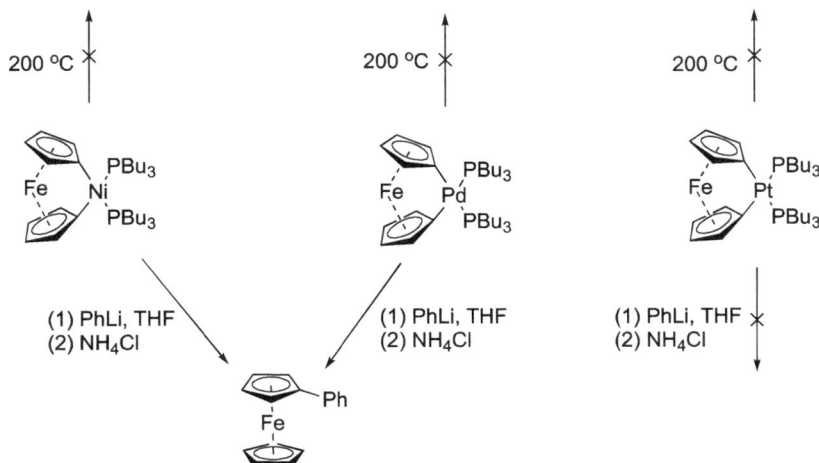

Figure 3.20 The chemistry of the organometallic molecule plays a pivotal role in its polymerisability. These group 10 metal containing [1]ferrocenophanes contain heteroatoms and have highly strained cyclic structures ($\alpha = 24.5$–$28.4°$) but are inert or too labile under thermal (200 °C) or anionic (PhLi, THF) conditions.

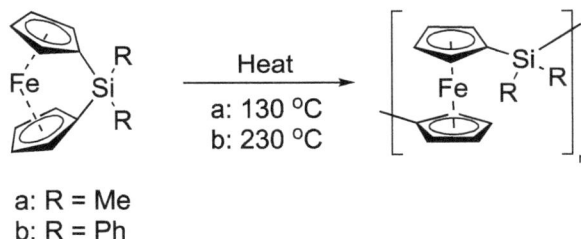

a: R = Me
b: R = Ph

Figure 3.21 Experimental conditions for ring-opening polymerisation of sila[1]ferrocenophanes depend on the substituent on the bridging heteroatom. Substituting a methyl with a phenyl changes the polymerisation temperature from 130 °C to 230 °C.

similar angle strain to tetracarba[4]nickelocenophane but lack torsional strain to undergo ROP (Box 3.7). The ROP of non-strained and moderately strained cyclic structures such as tetracarba[4]nickelocenophane and hexamethylcyclotrisiloxane implies that torsional strain contributes to the thermodynamic feasibility of the polymerisation.

3.5.1 Examples of ROP in Inorganic and Organometallic Polymer Synthesis

Ring-opening polymerisation is a well-established technique to synthesise inorganic and organometallic polymers with controlled molecular weight and dispersity. The technique allows the polymerisation of various cyclic molecules, including cyclosiloxanes, cyclosilanes, cyclophosphazenes, and [n]metallocenophanes. A typical ROP reaction consists of initiation, propagation, and termination steps. For most cationic ROPs, the initiator generates a cationic centre and, by the monomer's nucleophilic attack on the cation, it grows into a polymer (Figure 3.23a). On the other hand, anionic ROP involves forming an anionic centre, which attacks a monomer to grow the polymer chain (Figure 3.23b).

(a)

(b)

Pd(dba)$_2$, THF

40 °C, 48 hrs

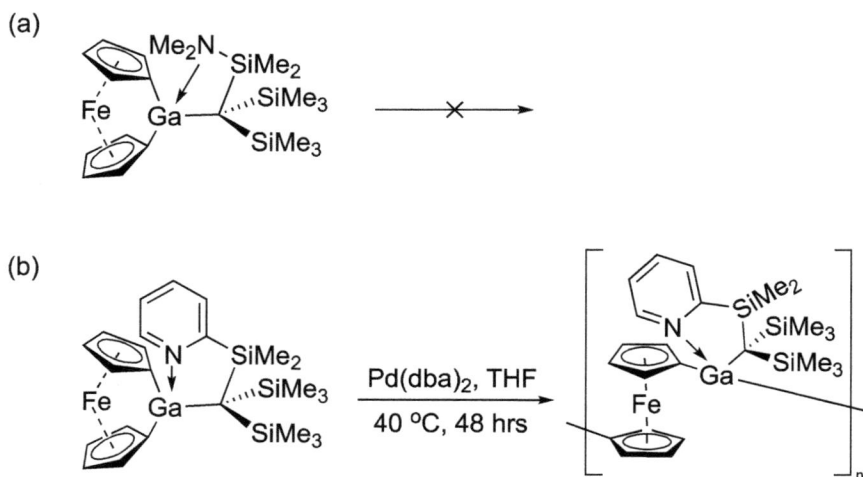

Figure 3.22 The electronic properties of the ligand coordinated to the bridging heteroatom influence the ROP of galla[1]ferrocenophanes. Substituting an electron-donating dimethylamino group (a) with an electron-withdrawing pyridyl group (b) results in ROP.

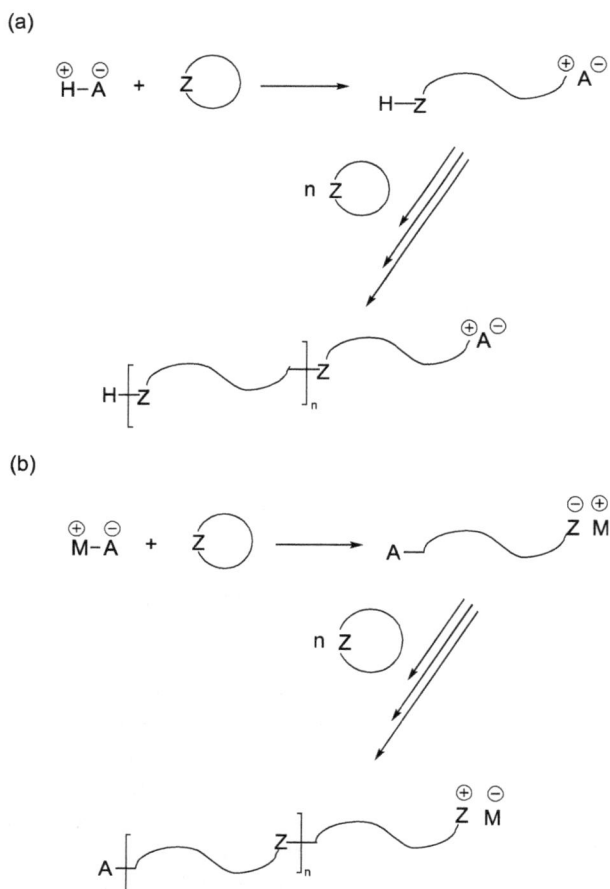

(a)

(b)

Figure 3.23 Schematic representation of the general mechanism of cationic (a) and anionic (b) ring-opening polymerisation.

The cationic ROP of hexamethylcyclotrisiloxane catalysed by a Brønsted–Lowry acid such as CF_3SO_3H is initiated by a proton attack on an oxygen atom, cleaving the ring to form a siliconium ion. Propagation involves a sequential nucleophilic attack by hexamethylcyclotrisiloxane molecules on the siliconium ion (Figure 3.24).[30] Lewis acids such as antimony pentachloride can co-initiate the cationic ROP of hexamethylcyclotrisiloxane with the hydrogen halide initiator to produce a polysiloxane with controlled molecular weight. This mechanism is similar to that of the thermal ROP of cyclophosphazenes, where the high temperature induces the ionisation of the phosphorus–chlorine bond, forming a phosphazenium ion. A nucleophilic attack by another cyclophosphazene on the phosphazenium ion propagates the reaction (Figure 3.25).[31] Anionic ROP requires basic initiators such as alkyllithium compounds or organic bases. Both cationic and anionic ROPs must be carried out at low temperatures to avoid the formation of unwanted side products. Transition metal-based initiators also induce the ROP of the cyclic structures, and in some cases, such as the ROP of cyclotetrasilane, these transition metal initiators are superior to alkyllithium initiators in controlling the polymer microstructure.

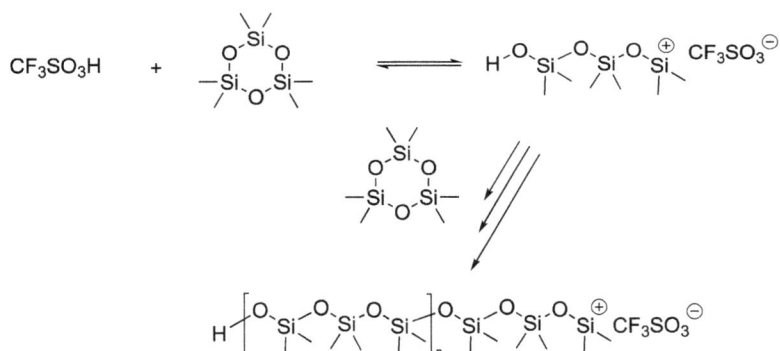

Figure 3.24 Mechanism of Brønsted–Lowry acid-catalysed cationic ring-opening polymerisation of hexamethylcyclotrisiloxane.

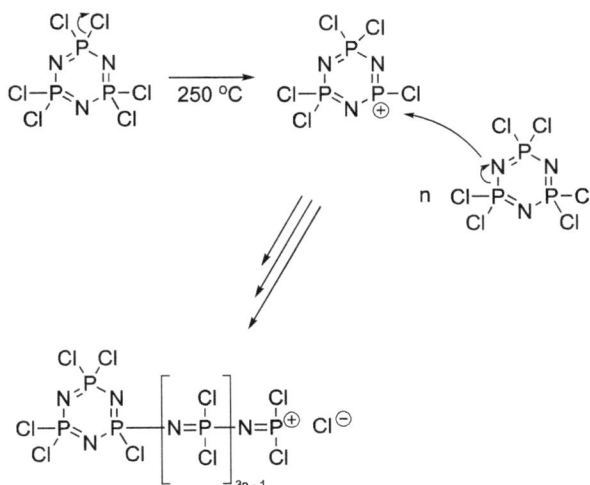

Figure 3.25 Mechanism of thermal-catalysed ring-opening polymerisation of cyclophosphazene.

(a)

(b)

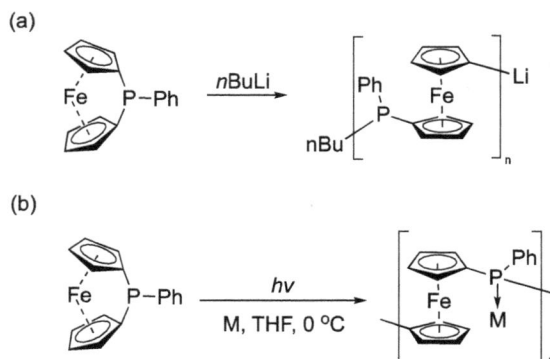

Figure 3.26 Synthesis of poly(ferrocenylphenylphosphine) *via* (a) anionic and (b) photolytic ring-opening polymerisation. M is a transition metal-based catalyst such as $[Mn(\eta\text{-}C_5H_4Me)(CO)_2]$, $[Mn(\eta\text{-}C_5H_5)(CO)_2]$, or $[W(CO)_5]$.

ROP of [n]metallocenophanes proceeds through various mechanisms, including thermal, anionic, transition metal catalysed, and photo-initiated polymerisation. The alkyllithium-initiated anionic ROP of phosphorus-bridged [1]ferrocenophane involves adding an alkyl group to the phosphorus atom, resulting in the cleavage of the phosphorus–cyclopentadienyl bond to create an anionic centre on the cyclopentadienyl ring for the propagation step (Figure 3.26). The photoinitiated ROP of phosphorus-bridged [1]ferrocenophane proceeds at 0 °C upon UV irradiation in polar solvents after a transition metal complex coordinates to the phosphorus atom.[25] In contrast, the photoinitiated ROP of the silicon-bridged [1]ferrocenophane entails complexation of the iron atom by solvent molecules, followed by nucleophilic attack by an initiator such as sodium cyclopentadienide, resulting in the cleavage of an iron–cyclopentadienyl bond (Figure 3.27).[25] Unlike photoinitiated ROP of the phosphorus-bridged [1]ferrocenophane, which is non-living and gives polymers with uncontrolled molecular weights, that of the silicon-bridged congener is living and yields well-defined polymers with controlled molecular weights.

3.6   Ring-opening Olefin Metathesis Polymerisation

Ring-opening olefin metathesis polymerisation (ROMP), a subclass of the ROP technique, is used to prepare polymers from cyclic olefins. The mechanism involves a metal-mediated scission and regeneration of carbon–carbon double bonds, resulting in the retention of the monomer's unsaturation in the polymer. The retention of the monomer's unsaturation in the polymer distinguishes ROMP from a typical chain polymerisation, where the unsaturated bond of the monomer is converted to a saturated bond. Most ROMP reactions use small molecule, strained cyclic olefins, such as norbornene and cyclopentene, as monomers with the thermodynamic driving force being derived from the ring strain. Conjugating organometallic moieties to these strained olefins, therefore, provides organometallic monomers. These monomers can be polymerised *via* ROMP using various heterogeneous and homogeneous catalysts;

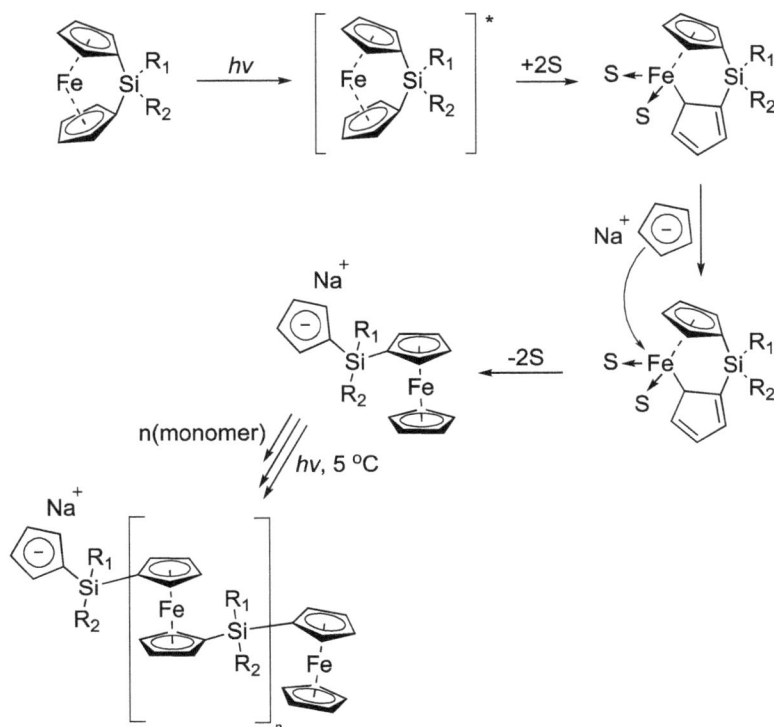

Figure 3.27 The mechanism of the photolytic anionic ring-opening polymerisation of silicon-bridged [1]ferrocenophane.

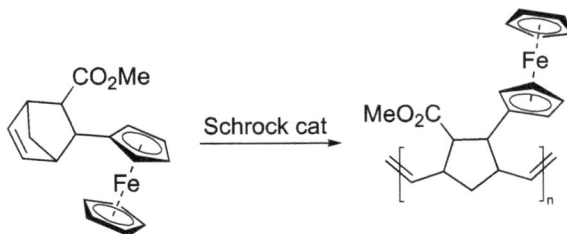

Figure 3.28 Ring-opening olefin metathesis polymerisation of the ferrocene-containing norbornene monomer. The ferrocene does not impact the living character of the polymerisation, as evidenced by the narrow molecular weight distribution ($Đ = 1.13$).

however, the Schrock and Grubbs catalysts are better suited for the living ROMP of these monomers. Generally, the presence of the organometallic group does not impact the living character of ROMP of norbornene monomers. For example, a ferrocene-containing norbornene (Figure 3.28) undergoes ROMP to produce polymers with narrow molecular weight distribution ($Đ = 1.13$),[32] which is within the range obtained under controlled polymerisation.

ROMP is a direct means to obtain well-defined homo- and copolymers bearing side-chain organometallic groups. For example, the norbornene conjugated to metal carbonyl (Figure 3.29) undergoes living ROMP, giving high molecular weight

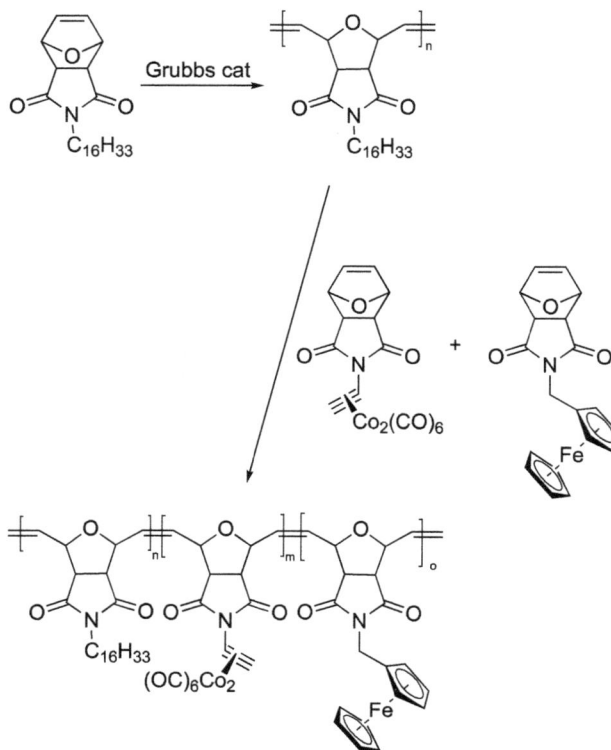

Figure 3.29 Synthesis of ferrocene- and cobalt-containing norbornene copolymers *via* ring-opening olefin metathesis polymerisation.

organometallic polymers.[33] The polymerisations are efficient, exhibiting a controlled mechanism and proceeding to nearly 100% monomer conversion in less than 15 minutes. The ferrocene- and cobalt hexacarbonyl containing norbornenes afford copolymers with monomodal and narrow molecular weight distribution ($Đ < 1.10$), implying a controlled polymerisation mechanism.[33] The organometallic group did not affect the kinetics of ROMP because of the similarity of the plot of ferrocene- and cobalt hexacarbonyl-containing norbornene conversion as a function of time (Figure 3.30). Isothermal controlled polymerisations are characteristically first-order kinetic reactions because the relative monomer composition remains constant with time. ROMP of these organometallic norbornene monomers follows the first-order kinetics, as evidenced by the linearity of the plot of $\ln[M]_0/[M]$ as a function of time.

Non-living ROMP of strain-free macrocyclic alkenes is emerging and provides a direct synthesis route to polymers with novel main-chain functionalities. A lack of strain means these polymerisations are non-enthalpy driven and must, therefore, be entropy-driven to bias the free energy towards polymer formation. Indeed, entropy-driven ROMP is feasible for concentrated solutions of high molecular weight monomers such as macrocycles considering that these monomers lose translational entropy upon polymerisation. Recently, a non-living ROMP of ferrocene-containing macrocycles composed of more than 11 atoms allows the synthesis of organometallic polymers bearing main-chain ferrocene (Figure 3.31). A high monomer concentration (200 mM to 800 mM) is required

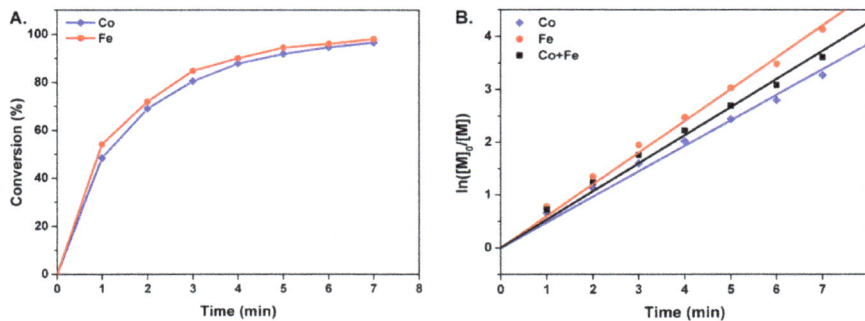

Figure 3.30 The organometallic groups did not affect the living character of the ring-opening olefin metathesis polymerisation. (A) A plot of monomer conversion as a function of time. (B) A plot of relative monomer composition ($\ln[M]_0/[M]$) as a function of time. Fe: ferrocene–norbornene monomer; Co: cobalt–norbornene monomer; Co + Fe: 1:1 mixture of Fe and Co. Reproduced from ref. 32 with permission from American Chemical Society, Copyright 2012.

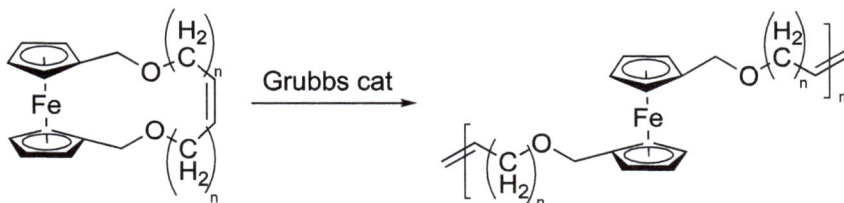

Figure 3.31 Ring-opening olefin metathesis polymerisation of ferrocene-containing macrocycle monomer for the synthesis of polymers containing main-chain ferrocene.

for successful polymerisation; therefore, the monomers should feature excellent solubility. Indeed, strain-free, macrocyclic monomers with limited solubility are unable to undergo non-living ROMP, while those exhibiting excellent solubility yield polymers with broad molecular weight distribution ($Đ > 1.50$), which is characteristic of non-living polymerisation (Box 3.9).[34]

Box 3.9 Review questions.

1. Consider the radical chain polymerisation of vinyl ruthenocene. Compared with vinylferrocene, will intramolecular chain transfer play a significant role in the termination step?
2. Describe how to design a cobaltocene-containing polymer, wherein cobalt is in the main chain of the polymer. Use equations.
3. Explain why NMP of ferrocenylmethyl methacrylate fails.
4. Consider the cationic polymerisation of ferrocenylmethyl methacrylate and methyl methacrylate. Predict which monomer will feature a higher polymerisation rate. Give reason(s).

Further Reading

1. J. E. Mark, H. R. Allcock and R. West, *Inorganic Polymers*, Oxford University Press Inc., New York, 2005.
2. R. D. Archer, *Inorganic and Organometallic Polymers*, Wiley-VCH, New York, 2001.
3. G. Odian, *Principles of Polymerization*, John Wiley & Son Inc, Hoboken, 2004.
4. A. S. Abd-El-Aziz, C. Agatemor and N. Etkin, *Macromol. Rapid Commun.*, 2014, **35**, 513–559.
5. C. U. Pittman Jr, C. E. Carraher Jr, M. Zeldin, J. E. Sheets and B. M. Culbertson, *Metal-containing Polymeric Materials*, Plenum Press, New York, 1996.
6. N. P. S. Chauhan and N. S. Chundawat, *Inorganic and Organometallic Polymers*, Walter de Gruyter GmbH, Berlin, 2019.

References

1. G. R. Buell, W. E. McEwen and J. Kleinberg, *J. Am. Chem. Soc.*, 1962, **84**, 40–43.
2. G. Buell, E. McEwen and J. Kleinberg, *Tetrahedron Lett.*, 1959, **1**, 16–18.
3. M. George and G. Hayes, *J. Polym. Sci. Polym. Chem.*, 1975, **13**, 1049–1070.
4. M. G. Baldwin and K. E. Johnson, *J. Polym. Sci., Part A: Polym. Chem.*, 1967, **5**, 2091–2098.
5. S. Chernyy, Z. Wang, J. J. K. Kirkensgaard, A. Bakke, K. Mortensen, S. Ndoni and K. Almdal, *J. Polym. Sci., Polym. Chem.*, 2017, **55**, 495–503.
6. P. W. Cyr, D. A. Rider, K. Kulbaba and I. Manners, *Macromolecules*, 2004, **37**, 3959–3961.
7. J. Ma, Y. Lin, Y.-W. Kim, Y. Ko, J. Kim, K. H. Oh, J.-Y. Sun, C. B. Gorman, M. A. Voinov and A. I. Smirnov, *ACS Macro Lett.*, 2019, **8**, 1522–1527.
8. A. Reyhani, H. Ranji-Burachaloo, T. G. McKenzie, Q. Fu and G. G. Qiao, *Macromolecules*, 2019, **52**, 3278–3287.
9. C. Yagci and U. Yildiz, *Eur. Polym. J.*, 2005, **41**, 177–184.
10. N. Zeng, Y. Yu, J. Chen, X. Meng, L. Peng, Y. Dan and L. Jiang, *Polym. Chem.*, 2018, **9**, 3215–3222.
11. R. Yao, R. Wu and G. Zhai, *Polym. Eng. Sci.*, 2015, **55**, 735–744.
12. J. L. Reddinger and J. R. Reynolds, *Synth. Met.*, 1997, **84**, 225–226.
13. R. Ren, Y. Wang, D. Liu and W. Sun, *Des. Monomers. Polym.*, 2017, **20**, 300–307.
14. P. Yang, P. Pageni, M. P. Kabir, T. Zhu and C. Tang, *ACS Macro Lett.*, 2016, **5**, 1293–1300.
15. T. Otsu and M. Yoshida, *Die Makromol. Chem. Rapid Commun*, 1982, **3**, 127–132.
16. M. K. Georges, R. P. Veregin, P. M. Kazmaier and G. K. Hamer, *Macromolecules*, 1993, **26**, 2987–2988.
17. M. Baumert, J. Fröhlich, M. Stieger, H. Frey, R. Mülhaupt and H. Plenio, *Macromol. Rapid Commun.*, 1999, **20**, 203–209.
18. C. U. Pittman Jr, J. Lai, D. Vanderpool, M. Good and R. Prado, *Macromolecules*, 1970, **3**, 746–754.
19. D. A. Poulsen, B. J. Kim, B. Ma, C. S. Zonte and J. M. Fréchet, *Adv. Mater.*, 2010, **22**, 77–82.
20. Y. Yan, J. Zhang, Y. Qiao, M. Ganewatta and C. Tang, *Macromolecules*, 2013, **46**, 8816–8823.
21. W. Qian, H. Zhang, T. Song, M. Ye, C. Feng, G. Lu and X. Huang, *Eur. Polym. J.*, 2019, **119**, 8–13.
22. C. Aso, T. Kunitake and T. Nakashima, *Die Makromol. Chem.*, 1969, **124**, 232–240.
23. O. Nuyken, V. Burkhardt and C. Hübsch, *Macromol. Chem. Phys.*, 1997, **198**, 3353–3363.
24. M. Gallei, B. V. Schmidt, R. Klein and M. Rehahn, *Macromol. Rapid Commun.*, 2009, **30**, 1463–1469.
25. R. Resendes, J. M. Nelson, A. Fischer, F. Jäkle, A. Bartole, A. J. Lough and I. Manners, *J. Am. Chem. Soc.*, 2001, **123**, 2116–2126.
26. I. Matas, G. R. Whittell, B. M. Partridge, J. P. Holland, M. F. Haddow, J. C. Green and I. Manners, *J. Am. Chem. Soc.*, 2010, **132**, 13279–13289.
27. S. O. Kang, J. W. Bausch, P. J. Carroll and L. G. Sneddon, *J. Am. Chem. Soc.*, 1992, **114**, 6248–6249.

28. J. A. Schachner, S. Tockner, C. L. Lund, J. W. Quail, M. Rehahn and J. Müller, *Organometallics*, 2007, **26**, 4658–4662.
29. R. A. Musgrave, R. L. Hailes, V. T. Annibale and I. Manners, *Chem. Sci.*, 2019, **10**, 9841–9852.
30. L. Wilczek, S. Rubinsztajn and J. Chojnowski, *Die Makromol. Chem.*, 1986, **187**, 39–51.
31. H. R. Allcock, D. J. Brennan and R. W. Allen, *Macromolecules*, 1985, **18**, 139–144.
32. Y. Zha, H. D. Thaker, R. R. Maddikeri, S. P. Gido, M. T. Tuominen and G. N. Tew, *J. Am. Chem. Soc.*, 2012, **134**, 14534–14541.
33. D. Albagli, G. Bazan, M. Wrighton and R. Schrock, *J. Am. Chem. Soc.*, 1992, **114**, 4150–4158.
34. Y. Sha, Y. Zhang, T. Zhu, S. Tan, Y. Cha, S. L. Craig and C. Tang, *Macromolecules*, 2018, **51**, 9131–9139.

4 Polymer Characterisation

4.1 Introduction

This chapter will discuss polymer characterisation, which involves using analytical methods and tests to determine the molecular mass, structure and morphology, and bulk macroscopic properties of inorganic and organometallic polymers. The characterisation required depends on the goal of the study, application of interest, and the availability and cost of the instrumentations involved. To explain, let us consider a hypothetical organometallic polymer with potential use as a biomaterial and engineering material. It would be highly relevant to conduct a cytotoxicity test to establish the biocompatibility of the polymer when used as a biomaterial. However, this test may be irrelevant if the polymer is intended as a precursor for fabricating aerospace materials. However, irrespective of application, it is pertinent to characterise the molecular weight because it is a fundamental property that influences other macroscopic properties and differentiates polymers from small molecules. Undoubtedly, a detailed and unambiguous characterisation requires combining various analytical methods and tests to unravel the molecular weight, structure and morphology, and macroscopic properties of interest (Box 4.1).

The field of polymer characterisation is vast, with different analytical methods and tests capable of providing information on the same polymer property. For instance,

---

Box 4.1 Learning outcomes.

By the end of this chapter, the student should be able to:

1. Discuss the importance of characterising polymers.
2. Evaluate different polymer characterisation methods.
3. Design a characterisation strategy to study a specific polymer property.
4. Extract information from characterisation data to predict the properties and applications of a polymer.

Fundamentals of Inorganic and Organometallic Polymer Science
By Christian Agatemor, Kajal Ghosal, Samuel Fura and Peter J. S. Foot
© Christian Agatemor, Kajal Ghosal, Samuel Fura and Peter J. S. Foot 2024
Published by the Royal Society of Chemistry, www.rsc.org

gel permeation chromatography (GPC) and matrix-assisted laser desorption ion-isation mass spectrometry (MALDI-MS) provide the average molecular weight (M_n and M_w) and molecular weight distribution ($Ð$) of polymers; therefore, choosing a method will depend on factors such as availability and operational cost. We must also note that characterising inorganic and organometallic polymers is more challenging than characterising organic polymers or small molecules. This challenge stems from the unique chemistry of the inorganic moieties in these polymers. For example, light scattering is a valuable tool to determine absolute molecular weight; however, unless we apply appropriate corrections, it is inapplicable to most organometallic and coordination polymers. The reason is that most organo-metallic and coordination polymers contain transition metals that absorb in the visible region used in light scattering measurements. Nevertheless, most methods used to characterise organic polymers apply to inorganic and organometallic polymers.

We must recognise that the scope of this monograph is insufficient to present ad-vanced principles and instrumentations of all the analytical methods discussed here. Therefore, interested students should consult advanced textbooks (please see the Further Reading section) for detailed discussions. Instead, we discuss fundamental principles with a focus on information relevant to polymer chemists. We centre the discussion around three subsections that emphasise polymer properties – molecular weight (Section 4.2), structure and morphology (Section 4.3), and bulk macroscopic properties (Section 4.4) – an approach that is more understandable to a polymer chemist. By the end of the discussion, we expect the student to grasp the fundamentals required to select an appropriate analytical method or test to evaluate the desired property.

4.2 Molecular Weight Characterisation

Molecular weight is a fundamental property of polymers. This property distinguishes polymers from small molecules and directly influences many macroscopic properties such as viscoelasticity and thermal stability. A hallmark of the molecular weights of synthetic polymers is non-uniform dispersity, which is due to the statistical variations inherent in the termination step of the polymerisation process. In a polymerisation system, the propagating chains terminate at different times, resulting in polymers with different molecular weights. A consequence of this statistical variation is that a polymer sample consists of chains of different molecular weights. Therefore, the ex-perimentally determined molecular weight is an average of the different molecular weights. To fully characterise polymer molecular weight, we must determine the average molecular weights and dispersity, which is the distribution of the different molecular weights.

Various analytical methods using polymer solutions allow the determination of the average molecular weights. These methods include light scattering measurement, vis-cosity measurement, end-group analysis, and the measurement of colligative properties (freezing point depression, boiling point elevation, osmotic pressure, and vapour pressure lowering). Each method assesses a unique property biased towards a subset of polymer chains, resulting in each method yielding different average molecular weight values. The two most common averages reported are the number-average (M_n) and the weight-average (M_w) molecular weights. Methods, such as end-group analysis and colligative property measurement, that evaluate the number of polymer chains or molar concentration of each specific weight in a sample yield the M_n. The number-average molecular weight is defined by eqn (4.1).

$$M_n = \frac{\sum N_i M_i}{\sum N_i} \tag{4.1}$$

N_i is the number of polymer chains or the number of moles of polymer chains with a specific molecular weight, M_i. Eqn (4.1) can be rewritten as eqn (4.2).

$$M_n = \frac{\sum W_i}{\sum N_i} \tag{4.2}$$

W_i is the weight of polymer chains with a specific molecular weight and N_i is the number of moles of chains with a particular molecular weight.

Rather than count the number of polymer chains, some analytical methods such as light scattering measure molecular weight from the weight of the chain with a specific molecular weight. The molecular weight obtained from these methods, known as the weight-average molecular weight (M_w), is defined by eqn (4.3).

$$M_w = \frac{\sum N_i M_i^2}{\sum N_i M_i} \tag{4.3}$$

N_i in eqn (4.1) has been replaced by $N_i M_i$, which is proportional to the weight (W_i) of polymer macromolecules with a specific molecular weight. Eqn (4.3) can also be written as eqn (4.4).

$$M_w = \frac{\sum W_i M_i}{\sum W_i} \tag{4.4}$$

M_w sums the weight fractions instead of the mole fractions and is mainly obtained using light scattering measurements. Because light scattering increases with molecular weight, the high molecular weight fraction of a polymer sample contributes more to M_w measurements (Box 4.2).

Box 4.2   Worked example 4.1.

Question
   Suppose a sample of polydimethylsiloxane contains 15 grams of polymer chains with a molecular weight of 50 000 g mol^{-1}, 5 grams of chains with 75 000 g mol^{-1}, and 5 grams of chains with 25 000 g mol^{-1}, compute the M_n and M_w. Also, comment on what biases M_w.

Answer

To calculate M_n, recall and apply eqn (4.2), which will yield the following expression:

$$M_n = \frac{15\,g + 5\,g + 5\,g}{\dfrac{15\,g}{50\,000\,g\,mol^{-1}} + \dfrac{5\,g}{75\,000\,g\,mol^{-1}} + \dfrac{5\,g}{25\,000\,g\,mol^{-1}}}$$

$$M_n = 44\,092\,g\,mol^{-1}$$

To calculate the M_w, recall eqn (4.4):

$$M_w = \frac{\left(15\,g \times 50\,000\,g\,mol^{-1}\right) + \left(5\,g \times 75\,000\,g\,mol^{-1}\right) + \left(5\,g \times 25\,000\,g\,mol^{-1}\right)}{15\,g + 5\,g + 5\,g}$$

$$M_w = 50\,000\,g\,mol^{-1}$$

The calculated M_w ($50\,000\,g\,mol^{-1}$) is higher than M_n because methods for estimating M_w are biased towards the large-sized macromolecules.

In light scattering, large-sized macromolecules, which more effectively scatter light than small-sized macromolecules, contribute more to M_w values. This bias towards large-sized macromolecules explains the higher value of M_w than that of M_n, except when the statistical variation in the termination step is eliminated to produce uniformly disperse polymers, an improbable occurrence in polymerisation. The ratio M_w/M_n, called dispersity ($Đ$), indicates the molecular weight distribution and is narrow ($Đ < 1.50$) under controlled polymerisation which significantly reduces termination reactions (Section 3.5).

Below, we discuss the methods used to determine absolute M_n and M_w from thermodynamic equations. Other average molecular weights such as viscosity-average molecular weight (M_v) are relative methods since they require calibration. Measurement of polymer solution viscosity gives M_v, defined by eqn (4.5).

$$M_v = \left(\frac{\sum N_i M_i^{1+a}}{\sum N_i M_i}\right)^{\frac{1}{a}} \tag{4.5}$$

a is a constant that quantifies the polymer–solvent interaction. When a is unity, M_w and M_v are equal, but for most polymers, a is in the range of 0.5–0.9, making M_v less than M_w but greater than M_n. Since solution viscosity is a hydrodynamic–thermodynamic property, the polymer–solvent interaction contributes to the value of M_v. The value of a, which characterises this interaction, depends on polymer hydrodynamic volume which varies with polymer molecular weight, solvent, and temperature.

4.2.1 Colligative Property Measurement

Colligative properties (the adjective is derived from the Latin word "*colligãtum*", which means "*connected*") are intensive properties of solutions. These properties depend on the number of solute particles in a solution rather than the identity of the solute. Analytical methods that measure colligative properties report the physical changes that result from adding a solute to a solvent. The freezing point, boiling point, vapour pressure and osmotic pressure of a solution are classic examples of colligative properties. Boiling point elevation, also known as ebulliometry, freezing point depression, also known as cryoscopy, vapour pressure lowering, and osmotic pressure enhancement can be used to determine M_n from well-established thermodynamic relationships. Usually, measurements are performed at different concentrations and extrapolated to infinite dilution, where the thermodynamic relationships are valid.

4.2.1.1 Cryoscopy and Ebulliometry

For a linear polymer with $Đ > 1$, M_n can be calculated using thermodynamic equations, which are similar for cryoscopy (eqn (4.6)) and ebulliometry (eqn (4.7)).

$$\left(\frac{\Delta T_f}{C}\right)_{c=0} = \frac{RT^2}{\rho \Delta H_f M_n} + A_2 C \tag{4.6}$$

$$\left(\frac{\Delta T_b}{C}\right)_{c=0} = \frac{RT^2}{\rho \Delta H_v M_n} + A_2 C \tag{4.7}$$

ΔT_f and ΔT_b are freezing point depression and boiling point elevation, respectively. ΔH_f and ΔH_v are the latent heats of fusion and evaporation per gram of solvent, respectively. C is the polymer concentration in grams per millilitre, T is the freezing point or boiling point of the solvent in Kelvin, R is the molar gas constant, ρ is the solvent density, and A_2 is the second virial coefficient. M_n is determined from a plot of $\Delta T_f/C$ or $\Delta T_b/C$ against C extrapolated to infinite dilutions (Figure 4.1).

Molecular weight determination using cryoscopy and ebulliometry requires thermally stable instruments sensitive to small temperature changes and can maintain temperature and concentration equilibrium. Indeed, temperature sensitivity is a limiting factor in ebulliometry and cryoscopy because as the molecular weight increases, temperature changes become smaller and difficult to measure, limiting the use of these methods to M_n of about 50 000 Da. The intrinsic insolubility of inorganic and

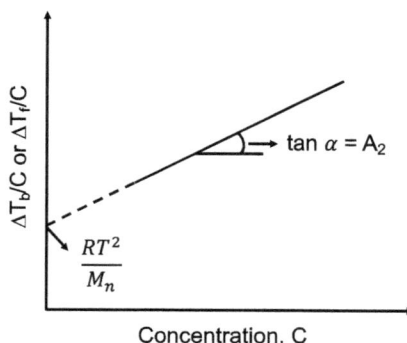

Figure 4.1 A plot of reduced boiling point elevation, $\Delta T_b/C$, or reduced freezing point depression, $\Delta T_f/C$, against concentration, C.

organometallic polymers can further limit the capability of the methods, which require polymer solutions for successful analysis. Improving solubility by converting the polymer from its crystalline to an amorphous form that is soluble in common laboratory solvents enables ebulliometric determination of the M_n of various cobalt(ɪɪ) and zinc(ɪɪ) phosphinate polymers and copolymers.[1]

4.2.1.2 Membrane Osmometry

Among colligative property-based methods, only membrane osmometry is sensitive enough to determine the M_n of high molecular weight, poorly soluble inorganic and organometallic polymers. The method depends on the chemical gradient that develops across a semipermeable membrane separating a pure solvent from its polymer solution (Figure 4.2). This gradient creates a net flow of solvent molecules from the solvent compartment through the semipermeable membrane to the solution compartment to produce a hydrostatic pressure. The hydrostatic pressure increases with the net flow until reaching equilibrium, where the solvent flow from each compartment is equal. At equilibrium, the hydrostatic pressure equals the osmotic pressure, which is related to the absolute molecular weight of the solute in the solution. For an ideal solution, where each component obeys Raoult's law, osmotic pressure (π) is related to molarity by eqn (4.8).

$$\pi = \frac{nRT}{V} \tag{4.8}$$

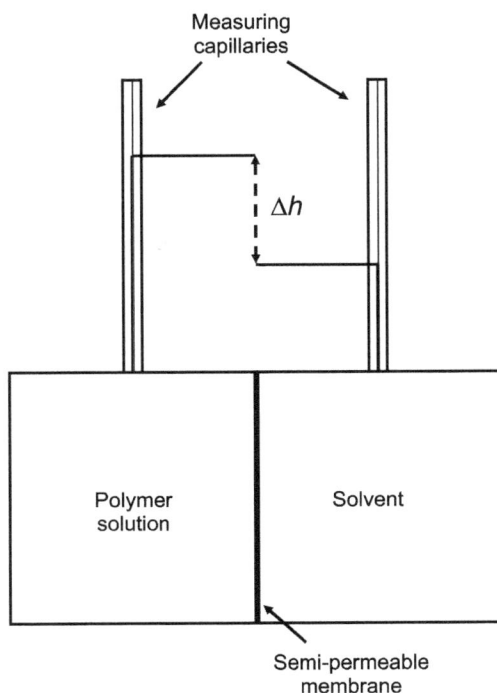

Figure 4.2 Basic principles of membrane osmometry.

n is the number of moles, V is the volume, and n/V is the molarity (m), which is equally defined by eqn (4.9).

$$m = \frac{C}{M} \qquad (4.9)$$

C is the concentration expressed in mass per volume and M is the molecular weight of the solute. Therefore, eqn (4.8) can be expressed as eqn (4.10).

$$\pi = \frac{CRT}{M} \qquad (4.10)$$

Because polymer solutions are nonideal solutions, it is required that π be measured at low polymer concentrations and extrapolated to infinite dilution. In this case, π is related to M_n by the thermodynamic relationship, eqn (4.11), derived for an infinitely dilute solution.

$$\left(\frac{\pi}{C}\right)_{C=0} = \frac{RT}{M_n} + A_2 C \qquad (4.11)$$

For membrane osmometry, π, in eqn (4.11), is defined by eqn (4.12).

$$\pi = \rho g \Delta h \qquad (4.12)$$

ρ is the density of the solvent in grams per millilitre, g is the acceleration due to gravity ($9.81~\mathrm{m\,s^{-2}}$), and Δh is the difference between the heights of the solvent and solution expressed in centimetres (Figure 4.2). In eqn (4.11), π/C is reduced osmotic pressure in $\mathrm{J\,kg^{-1}}$, R is the molar gas constant ($8.314~\mathrm{J\,mol^{-1}\,K^{-1}}$), T is the temperature in Kelvin, M_n is the number-average molecular weight of the polymer, C is the concentration in grams per litre, and A_2 is the second virial coefficient which provides a quantitative measure of the polymer–solvent interaction. The larger the value of A_2 is, the more the polymer solution deviates from the ideal, implying a stronger polymer–solvent interaction.

Usually, M_n is obtained from a plot of π/C against C (Figure 4.3). Multiple measurements of π are performed to extrapolate to infinite dilution. The plot is linear with the slope and intercepts equal to A_2 and RT/M_n, respectively. If measurements are made

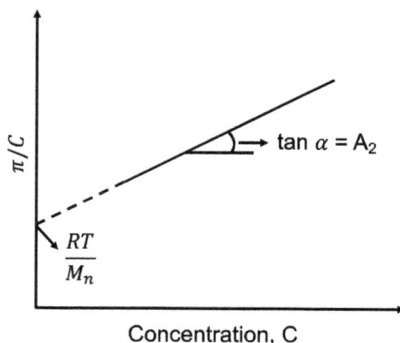

Figure 4.3   A plot of reduced osmotic pressure, π/C, against concentration, C.

Box 4.3 Worked example 4.2.

Question 4.2a

The M_n determined by membrane osmometry is higher than that obtained using GPC. Explain.

Answer

In membrane osmometry, the low-molecular-weight polymer diffuses through the membrane, leaving only a high molecular weight fraction for analysis.

Question 4.2b

Membrane osmometry obtains higher M_n values for step polymers than those for addition polymers. Explain.

Answer

Compared with addition polymerisation, step polymerisation yields a sizeable amount of low molecular weight polymers, many of which diffuses through the membrane leaving only the high molecular weight fraction for analysis.

under theta conditions, where the polymer–solvent interaction is negligible and the polymer–polymer interaction is preferred, A_2 is zero and π/C is equal to RT/M_n. Dynamic and static osmometers can measure π, with the former generating results within 10 minutes compared with the several hours in the latter.

The M_n determined by membrane osmometry is generally higher than the values obtained with other colligative property-based methods. This discrepancy arises because lower molecular weight polymers pass through the semipermeable membrane, resulting in only high molecular weight polymers being measured. This problem can be circumvented by precipitating out the high molecular weight polymer fractions for characterisation. Nevertheless, the practical limit of membrane osmometry covers the molecular weight range of most inorganic and organometallic polymers. The method successfully determines the M_n of polysiloxane $(M_n = 46\,920–91\,680 \text{ Da})^2$ and polyphosphazene $(M_n = 310\,000–1\,800\,000 \text{ Da})^3$ (Box 4.3).

4.2.2 End Group Analysis

Chemical analysis of the functional groups or structures attached to the periphery of linear or branched polymers furnishes data to estimate M_n. The analysis relies on the finding that the chemical reactivity of functional groups is independent of molecular weight, provided the polymer is of moderate molecular weight and sufficiently soluble to preclude a diffusion-controlled reaction. Because the relative concentration of end groups decreases with an increase in molecular weight, the practical utility of end group analysis is limited to 50 000 Da. For inorganic and organometallic polymers, the end group could be a radioactive initiator fragment that is detectable *via* radiochemical analysis or a functional group/structural motif that is detectable using volumetry or spectroscopy. Non-aqueous potentiometry is a widely used end group analytical method for measuring the M_n of polymers, but with inorganic and organometallic polymers, the sensitivity could be low due to their poor solubility in common laboratory solvents.

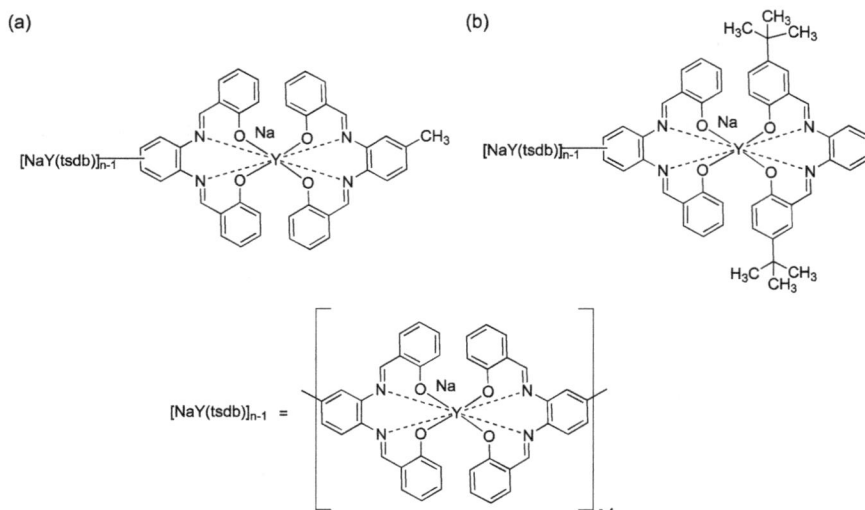

Figure 4.4 A design principle to enhance the sensitivity of ^{1}H NMR-based end group analysis by changing the end cap from a methyl (a) to a *tert*-butyl group (b).

Proton nuclear magnetic resonance (^{1}H NMR) spectroscopy is widely used, but its sensitivity may be limited since the polymer main chain signal could mask those from the end groups. However, this problem can be solved by modifying the end group to amplify the signal for improved sensitivity and ultimately increase the accuracy of the analysis. For example, in salen polymers (Figure 4.4), changing the end group from a methyl group in N,N'-bis(salicylidene)-1,2-diamino-4-toluene[4] to *tert*-butyl in N,N'-bis(4-*tert*-butylsalicylidene)-1,2-diaminobenzidine[5] improves the accuracy of a ^{1}H NMR-based end group analysis. The 18 protons in the 4-*tert*-butyl group in N,N'-bis(4-*tert*-butylsalicylidene)-1,2-diaminobenzidine is more ^{1}H NMR sensitive than the three protons in the toluene group of N,N'-bis(salicylidene)-1,2-diamino-4-toluene. As a result, the M_n values obtained with salen polymers end-capped with N,N'-bis(4-*tert*-butylsalicylidene)-1,2-diaminobenzidine were closer to the GPC determined M_n than those obtained with polymers having an N,N'-bis(salicylidene)-1,2-diamino-4-toluene end group (Box 4.4).

Box 4.4 Worked example 4.3.

Question

Consider the polymer poly(ferrocenyldimethylsilane) end-capped by ethylene sulphide.[6] The ^{1}H NMR shows that the two protons of the methylene group attached to the sulphur end group integrate as 0.0593, while the eight protons of the cyclopentadienyl group per repeat unit integrate as 5.4720; estimate the number of repeat units.

Answer

$$\text{Degree of polymerisation} = \frac{5.4720}{8} \Big/ \frac{0.0593}{2} = 23$$

4.2.3 Mass Spectrometry

Mass spectrometry is a powerful method to determine the absolute molecular weight of polymers. In a typical mass spectrometer, the sample is ionised, separated according to their mass–charge ratio (m/z), and counted in a detector. Soft ionisation techniques, such as electrospray ionisation (ESI) and matrix-assisted laser desorption/ionisation (MALDI), that generate mainly parent ions are mostly used in the characterisation of the M_n, M_w, and $Đ$ of polymers. For example, in MALDI mass spectroscopy, the polymer sample is embedded in a low molecular weight organic compound matrix, such as *trans*-2-[3-(4-*tert*-butylphenyl)-2-methyl-2-propenylidene]malononitrile, that absorbs in the wavelength region of an ultraviolet laser. Because synthetic polymers are difficult to ionise by protonation, adding a cationising agent such as sodium, potassium silver, or copper salts to the matrix is necessary. For inorganic and organometallic polymers, the presence of the heteroatoms enhances the ionisation efficiency, so sodium or potassium salt will suffice as a cationising agent. Upon irradiation, the matrix transfers the laser energy to the polymer, which desorbs (vapourise) to form polymer molecules with attached metal ions. For polymers with narrow molecular weight distribution ($Đ < 1.3$), MALDI-MS provides accurate molecular weight distribution values (M_n/M_w) which agree with those obtained with gel permeation chromatography. The M_n and M_w values are computed using eqn (4.13) and (4.14).

$$M_n = \frac{\sum N_i M_i}{\sum N_i} \tag{4.13}$$

$$M_w = \frac{\sum N_i M_i^2}{\sum N_i M_i} \tag{4.14}$$

N_i is the number of ions of the ith-mer, obtained from the integrated peak areas, and M_i is the molecular weight of the ith-mer.

MALDI-MS is a valuable method to characterise the molecular weight of cationic organometallic polymers, which are challenging to determine using widely used gel permeation chromatography (Section 4.2.6). For example, the method characterises the molecular weight of the polycationic copolymer (Figure 4.5a) containing ferrocene and cobaltocenium.[7] A typical MALDI-MS mass spectrum is shown in Figure 4.5b. Figure 4.5b shows a segment of the mass spectrum for the copolymer in Figure 4.5a, and it shows well-defined peaks of fragments separated by the molecular weight difference that corresponds to the monomers' molecular weight.

4.2.4 Light Scattering

This technique relies on the principle that electromagnetic radiation such as visible light loses energy by absorption or scattering on passing through a liquid medium. For polymer characterisation, a light scattering experimental setup consists of a source of a well-collimated, monochromatic, polarised light beam usually obtained with a laser, a trap for the non-scattered incident beam, and a turntable-mounted photocell which measures the intensity of scattered light as a function of scattering angle, θ, and

(a)   (b)

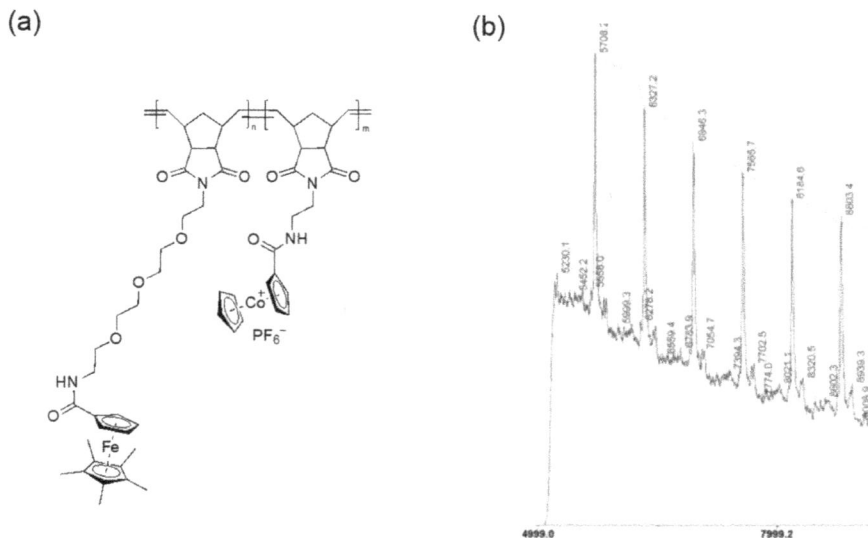

Figure 4.5   (a) Structure of the cationic metal-containing copolymer characterised by MALDI-MS. (b) Mass spectrum of the copolymer showing peaks separated by the mass difference that corresponds to the molecular mass of monomers. Adapted from ref. 7 with permission from American Chemical Society, Copyright 2016.

concentration, c. The electric field component of the incident light should be generated perpendicularly to the plane used to measure the intensity and angular dependence of the scattered light. The intensity of the scattered light, which depends on the polarisability and concentration of the scattering macromolecules, contains information on molecular weight. In contrast, the angular dependence of the scattered light measured in the horizontal plane carries information on the size of the polymer.

A significant quantity is the Rayleigh ratio (R_θ) (eqn (4.15)), which is the fraction of the incident light scattered at angle θ.

$$R_\theta = \frac{i_\theta r^2}{I_0} \tag{4.15}$$

i_θ is the intensity of the scattered light, I_0 is the intensity of incident light, and r is the distance from the scattering cell to the detector.

We must consider the angular dependence of the scattered light intensity and solvent and polymer molecule contributions to light scattering. The contribution from the solvent is simply corrected as defined in eqn (4.16).

$$R_{\theta\text{polymer}} = R_\theta - R_{\theta\text{solvent}} \tag{4.16}$$

The angular dependence of the scattered light intensity is corrected for by including two factors: $\sin \theta$, to account for the increase in scattering intensity as the scattering angle deviates from the reference perpendicular angle, and $1/(1 + \cos^2 \theta)$ to account for the difference in the horizontal and vertical components of the scattered light. Incorporating these two corrections yields a new Rayleigh ratio, R'_θ, defined by eqn (4.17). The quantity

Kc/R_θ, known as the scattering function, relates R_θ' with concentration, c. The quantity K is the optical constant in the scattering function and is defined by eqn (4.18).

$$R_\theta' = R_{\theta\text{polymer}}\left[\frac{\sin\theta}{(1+\cos^2\theta)}\right] \tag{4.17}$$

$$K = \left(\frac{2\pi^2 n_0^2}{N\lambda^4}\right)\left(\frac{\mathrm{d}n}{\mathrm{d}c}\right)^2 \tag{4.18}$$

n_0 is the refractive index of the solvent, N is Avogadro's number, λ is the wavelength of the incident light, and $\mathrm{d}n/\mathrm{d}c$ is the specific refractive increment obtained from the slope of a plot of the refractive index of the polymer solution as a function of its concentration. $\mathrm{d}n/\mathrm{d}c$ is constant for a given polymer solution at a given temperature.

A typical light scattering experiment for molecular weight determination is carried out at different concentrations with the intensity of the scattered light measured at different angles. Low polymer concentrations are required to mitigate interparticle interactions that could influence result interpretation. Extrapolating the concentration to zero reduces interparticle interactions, but as the molecular weight increases, the intraparticle interaction becomes relevant, necessitating extrapolation of both the concentration and scattering angle to zero. Plotting the values of the scattering function at zero scattering angle against concentration yields a graph with an intercept that equals the reciprocal of molecular weight. Conversely, plotting the scattering function at zero concentration gives a graph with a slope that equals the radius of gyration, which is related to polymer size. The extrapolation to zero concentration and zero scattering angle is done concurrently to obtain a Zimm plot (Figure 4.6),[8] which is a plot of the scattering function, Kc/R_θ, against $kc + \sin^2(\theta/2)$, where k is an arbitrary constant that determines how spread the data will be in the plot.

Although light scattering is widely used to determine the absolute M_w of polymers, it has limited use for inorganic and organometallic polymers that absorb in the visible region, where the scattered light is measured. In this case, corrections must be made to account for light absorption. Accurate scattering results also depend on clear, particle-free solutions, which are obtained by filtration or centrifugation. Despite these

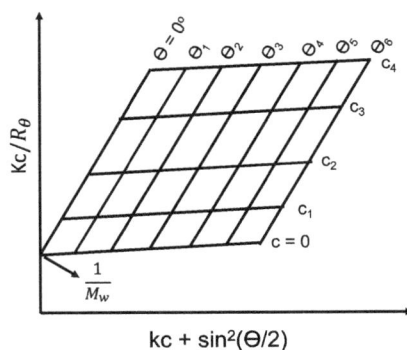

Figure 4.6   A typical Zimm plot showing the simultaneous extrapolation to zero scattering angle and zero concentration to determine the intercept, which equals the reciprocal of molecular weight.

challenges, the method has been used to estimate the molecular weight and size of many inorganic polymers, including polyphosphazenes, polygermanes, and polysilanes. The molecular weights obtained with light scattering range from 5000 to 10 000 000 Da and differ from those obtained with gel permeation chromatography using polystyrene standards.

Another scattering technique that can circumvent the issue of light absorption, small-angle X-ray scattering (SAXS), is also used to characterise polymer morphology (Section 4.3.3). Although less commonly available, SAXS is based on the deflection of a collimated incident X-ray beam and relies on the intrinsic difference in electron density between the solvent and the polymer solute. The scattering data obtained from SAXS and fitted to a Debye model for a monodisperse chain in a theta solvent allow the determination of the M_w of a polyvanadoarenophane in THF.[9] A caveat, however, is that the polymer must be sufficiently soluble in the solvent for accurate SAXS characterisation of the molecular weight, a condition that most high molecular weight coordination and organometallic polymers fail to meet.

4.2.5 Viscometry

Viscosity measurement is an inexpensive method extensively used to determine the molecular weights of inorganic and organometallic polymers, including ferrocene-containing acrylate and methacrylate polymers.[10,11] The method gives the viscosity-average molecular weight (M_v), whose value lies between the M_w and M_n of the same polymer. Unlike absolute methods such as light scattering and mass spectrometry, viscometry gives relative molecular weights, relying on a calibration curve to obtain the quantities used to compute the molecular weights. The method uses Poiseuille's equation (eqn (4.19)) to obtain the viscosity coefficient (η) of the polymer solution from the flow time (t) it takes for a given volume (V) of the solution to flow through a capillary of radius (r) and length (L) under a differential pressure (p) at a specific temperature.

$$\eta = \frac{\pi p r^4 t}{8LV} \tag{4.19}$$

A classic viscometry experiment requires a clear, particle-free solution because particles affect flow time. For molecular weight determination, the viscosity ratio (η_{rel}) (eqn (4.20)), which is the ratio of the polymer solution viscosity to the solvent viscosity at a given temperature, provides the required information to compute the molecular weights. To a first approximation for dilute solutions, η_{rel} is proportional to the ratio of the flow times, which is the only quantity that varies significantly during the measurement.

$$\eta_{rel} = \frac{\eta_{solution}}{\eta_{solvent}} = \frac{t_{solution}}{t_{solvent}} \tag{4.20}$$

A specific viscosity parameter (η_{sp}), defined by eqn (4.21), is the fractional increase in viscosity due to an increase in polymer concentration. To eliminate the effects of concentration, η_{sp} is divided by the concentration and extrapolated to zero concentration to obtain the limiting viscosity number ([η]) (eqn (4.22)). The limiting viscosity number is related to M_v by the Mark–Houwink–Sakurada equation (eqn (4.23)). M_v is defined by eqn (4.24).

$$\eta_{sp} = \frac{\eta_{solution} - \eta_{solvent}}{\eta_{solvent}} = \frac{t_{solution} - t_{solvent}}{t_{solution}} = \eta_{rel} - 1 \qquad (4.21)$$

$$[\eta] = \left(\frac{\eta_{sp}}{C}\right)_{c=0} \qquad (4.22)$$

$$[\eta] = KM_v^a \qquad (4.23)$$

$$M_v = \left(\frac{\sum N_i M_i^{1+a}}{\sum N_i M_i}\right)^{1/a} \qquad (4.24)$$

K and a are the Mark–Houwink–Sakurada constants that depend on the polymer, solvent, and temperature.

The Mark–Houwink–Sakurada constants are evaluated from a plot of $\log[\eta]$ against $\log M$ with M being the M_w or M_n of a series of fractionated polymer samples (Figure 4.7). For linear polymers with narrow molecular weight distribution, these plots are linear and give slope and intercept corresponding to $\log K$ and a, respectively (eqn (4.25)), except at low molecular weights. Several factors such as polymer branching, monomer sequence, and polymer adsorption on capillary walls affect M_v determined from viscometry.

$$\log[\eta] = \log K + a \log M \qquad (4.25)$$

4.2.6 Gel Permeation Chromatography

Gel permeation chromatography is the most widely used method for routine determination of molecular weights and molecular weight distributions. It is a liquid chromatography technique (LC) that uses a liquid mobile phase and a solid stationary phase to separate a polymer sample into narrow molecular weight fractions. However, it differs from conventional LC techniques in its separation mechanism that relies

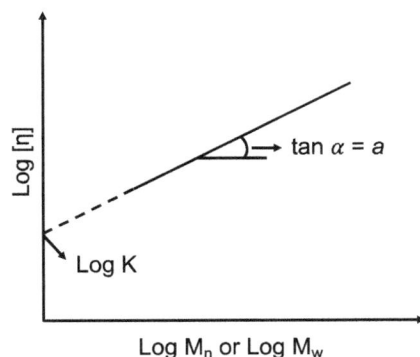

Figure 4.7 A plot of $\log[\eta]$ against $\log M_n$ or $\log M_w$ for determining the Mark–Houwink–Sakurada constants for a given polymer.

exclusively on differences in the hydrodynamic volume of polymer fractions. Separation occurs inside a column packed with the stationary phase, which is made up of a crosslinked gel such as crosslinked polystyrene–divinylbenzene porous beads. The essential components of a typical gel permeation chromatography (GPC) instrument include a solvent reservoir to supply the mobile phase, a column to package the stationary phase that separates the polymers, a pump to push the mobile phase through the instrument and a detector to detect the eluting polymer fractions (Figure 4.8).

A critical step in GPC analysis is to solubilise the polymer in a solvent that is compatible with the column. This step could challenge the characterisation of inorganic and organometallic polymers, which are primarily insoluble in common laboratory solvents. Fortunately, columns such as the commercially available Waters HSPgel RT series tolerate various solvents, including dimethyl formamide, dimethyl acetamide, N-methylpyrrolidone, dimethyl sulfoxide, and hexafluoroisopropanol, that dissolve several organometallic polymers, enabling GPC characterisation of these polymers. Also, the polymer must not chemically interact with the column because such interactions could degrade the polymer, preclude elution, or distort the GPC data. Unfortunately, some cationic organometallic polymers, such as the cobaltocenium-containing polymer in Figure 4.5, interact with some columns, precluding their GPC characterisation. Once the polymer fractions elute the column, a detector such as a refractive index detector, an ultraviolet detector, a photodiode array detector, a refractive index detector coupled to a light scattering detector or a refractive index detector coupled to a viscometer and a light scattering detector is used to detect the concentration or obtain other information about the eluted polymer. The results are presented as a chromatogram (Figure 4.9).

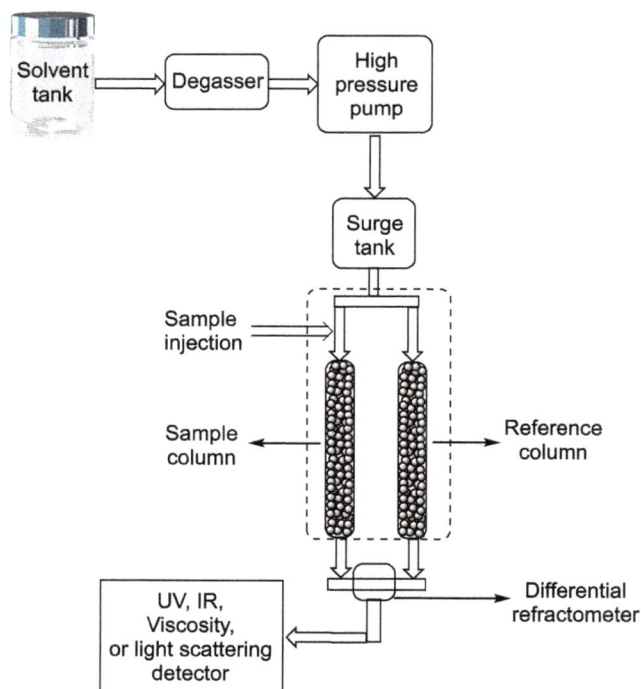

Figure 4.8 Critical components of gel permeation chromatography.

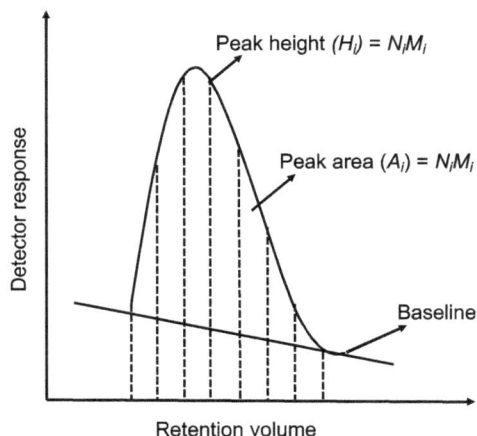

Figure 4.9 A typical GPC chromatogram partitioned into equally spaced "slices" of specific retention volumes with the area (A_i) or height (H_i) of each slice being proportional to the amount of polymer (N_iM_i).

A GPC chromatogram is a plot of detector response against retention volume, which can be converted to retention time, provided the flow rate is constant (Figure 4.9). The chromatogram displays the amount of each polymer fraction (N_iM_i) eluting the column at each time, with the highest molecular weight fraction eluting first, followed by consecutively lower molecular weight fractions. The separation mechanism is based on the inability of the high molecular weight polymers to penetrate the pores of the stationary phase and therefore have a shorter residence time in the pore compared with low molecular weight polymers that penetrate the pore and have a longer residence time. The overall effect is that high molecular weight polymers travel faster with the mobile phase and are eluted before low molecular weight fractions.

It is possible to obtain the absolute average molecular weight using multiple detectors such as a refractive index detector coupled to a viscometer and a light scattering detector. This multi-detector platform provides other helpful information such as the intrinsic viscosity, degree of branching and radius of gyration about the sample. The capability of determining the absolute molecular weight is beneficial for inorganic and organometallic polymers that lack an appropriate calibration standard required for relative molecular weight characterisation. Determining relative average molecular weights at a given retention time requires the chromatogram to be calibrated against a reference chromatogram obtained with a polymer of a known average molecular weight and narrow molecular weight distribution ($Đ < 1.10$). The GPC of a series of standards is run and a polynomial fit is performed to obtain a calibration curve of a plot of $\log M$ against retention volume or time. The calibration curve can then be used to determine the molecular weight of the sample by simply reading the molecular weight from the curve with the retention time of the sample. To obtain M_n, M_w and $Đ$, the chromatographic peak is divided into several equally spaced "slices" of specific retention time or volume with the area (A_i) or height (H_i) of each slice being proportional to the amount of polymer (N_iM_i). The molecular weight (M_i) of the "slice" is read from the calibration curve with the retention time. The molecular weight distribution $(Đ)$ is computed from M_n and M_w defined by eqn (4.26)–(4.28) (Box 4.5).

Box 4.5 Worked example 4.4.

Question

  GPC was used to separate a polymer sample into three fractions containing 60 polymer chains of mass 150 000 Da, 50 polymer chains of mass 100 000 Da, and 50 polymer chains of mass 50 000 Da. Compute the M_n, M_w, and Đ of a polymer.

Answer

  Using eqn (4.27) and (4.28)

$$M_n = \frac{(50 \times 100\,000) + (50 \times 50\,000) + (60 \times 150\,000)}{50 + 50 + 60} = 103\,125 \;\; Da$$

$$M_w = \frac{(50 \times 100\,000^2) + (50 \times 50\,000^2) + (60 \times 150\,000^2)}{(50 \times 100\,000) + (50 \times 50\,000) + (60 \times 150\,000)} = 119\,697 \;\; Da$$

$$Đ = M_w/M_n = 1.16$$

$$\sum A_i = \sum N_i M_i \tag{4.26}$$

$$M_n = \frac{\sum N_i M_i}{\sum N_i} = \frac{\sum A_i}{\sum A_i/M_i} \tag{4.27}$$

$$M_w = \frac{\sum N_i M_i^2}{\sum N_i M_i} = \frac{\sum A_i M_i}{A_i} \tag{4.28}$$

4.2.7 Diffusion-ordered NMR Spectroscopy

This is an emerging method to characterise the molecular weight, ranging from 100 to 1 000 000 Da, and molecular weight distribution. A diffusion-ordered NMR spectroscopy, DOSY, experiment uses a pulsed-field gradient spin-echo to obtain the translational self-diffusion coefficient, D, of polymers in solution. The Stokes–Einstein law (eqn (4.29)) relates D to the molecular size, specifically the hydrodynamic radius, R_H, of the self-diffusing polymer.

$$D = \frac{kT}{6\pi\eta R_H} \tag{4.29}$$

k is the Boltzmann constant, T is the absolute temperature, and η is the solution viscosity coefficient.

  Although the Stokes–Einstein law does not relate D to molecular mass, M, for spherical particles, M can be estimated from R_H using eqn (4.30).

$$R_H = \left(\frac{3M}{4\pi\rho N}\right)^{1/3} \tag{4.30}$$

ρ is the density of the solution and N is Avogadro's number.

It must, however, be noted that deviation from eqn (4.30) occurs when the actual shape of the molecule is non-spherical. For molecules about the same shape and approximately similar ρ, various DOSY studies relate D to M through the power-law relationship (eqn (4.31)).[12]

$$M \approx \left(\frac{C}{D}\right)^{d_{\mathrm{f}}} \tag{4.31}$$

d_{f} is the fractal dimension of the molecules and C is the calibration constant. Both d_{f} and C are characteristic of a family of molecules and must be determined experimentally. Linearising eqn (4.31) by taking the logarithm on both sides and making $\log D$ the subject of the equation yields eqn (4.32).

$$\log D = -\frac{1}{d_{\mathrm{f}}} \log M + \log C \tag{4.32}$$

Combining eqn (4.30)–(4.32) yields eqn (4.33).[13]

$$\log D = -\frac{1}{3} \log M + \frac{1}{3} \log \rho - \log \eta - \frac{1}{3} \log \frac{162\pi^2}{k^3 T^3 N_{\mathrm{A}}} \tag{4.33}$$

Provided that convection, density, and viscosity are consistent, a plot of $\log D$ against $\log M$ will be linear for dilute solutions. From a linear calibration curve generated from polymer standards with known molecular weights and narrow molecular weight distributions, the M_{w} of unknown polymers can be estimated with DOSY measured diffusion coefficients. DOSY has been used to characterise the M_{w} of polyoxometalates[12] and the R_{H} of iron-containing dendrimers.[14] The advantage of the method relative to GPC is the low operational cost. Specifically, DOSY experiments do not require large volumes of solvents and expensive columns. The experiment works with polymer solutions in a heterogeneous medium such as a turbid suspension, provided the signal-to-noise ratio is high (>500). A successful DOSY experiment requires controlled convection, long relaxation times independent of molecular weight, and low sample concentrations to ensure consistent viscosity and density.

4.3 Structural and Morphological Characterisations

In Section 4.2, we discussed how to characterise the molecular weight, a fundamental property of polymers. However, to fully describe a polymer, we must elucidate the chemical composition, macrostructure, microstructure, and morphology. A compositional analysis allows us to identify the constituent elements in a polymer, providing information to distinguish between organic, inorganic, and organometallic polymers. On the other hand, structural characterisation unravels the sequence and three-dimensional arrangement of monomers, while morphological analysis describes the microscale ordering of polymer chains. Structural and morphological characterisation is instrumental to a detailed understanding of polymers because these characterisations, alongside molecular weight information, furnish information to predict macroscopic properties, which will be discussed in Section 4.4. This section will focus on the various spectroscopic and microscopic tools widely used and readily available to characterise the composition, structure, and morphology of inorganic and organo-metallic polymers.

4.3.1 Nuclear Magnetic Resonance Spectroscopy

Section 4.2 focuses on NMR as a method to determine M_n through DOSY and end-group analysis. This section will discuss NMR as a routine analytical tool to identify the composition and elucidate the structure of polymers. The pre-eminent use of NMR spectroscopy in polymer characterisation is based, in part, on the observation that NMR peaks are assignable to specific atoms in the polymer network. The method provides atomic-scale qualitative and quantitative information on the functional groups, topology, conformation, stereochemistry, molecular dynamics, and three-dimensional structure of solution- and solid-state polymers. For example, in ^{1}H NMR, the area under the peak correlates with the number of spins; therefore, peak integrals can quantitatively determine the number of protons in the chemical environment.

We assume that the student is familiar with the fundamental principles of NMR spectroscopy. Nevertheless, we will recap a few basics to help the student decide if NMR is suitable for characterising inorganic and organometallic polymers (Box 4.6). Unlike organic polymers, where proton (^{1}H) and carbon-13 (^{13}C) are the most routinely studied NMR nuclei, inorganic and organometallic polymers contain other nuclei that may be NMR active. The main criterion for NMR activity is that the nucleus possesses a value of

---

Box 4.6   What influences the ease of observing an NMR signal?

Most routine NMR experiments involve ^{1}H and ^{13}C. Experiments with most inorganic elements, which are components of inorganic and organometallic polymers, are also typical. Therefore, it is essential to understand the factors that influence the ease of observing an NMR signal. The ease of observing an NMR signal depends on recep-tivity, chemical shift range, and linewidth. The receptivity of a nucleus directly cor-relates with the magnetogyric ratio (γ) and natural abundance, which are intrinsic properties of the nucleus. It is important to note that a high γ alone does not assure an observable NMR signal and likewise a high natural abundance. For example, ^{3}H has the highest γ but its extremely low natural abundance makes the NMR signal unobservable unless enriched. Also, ^{103}Rh with 100% natural abundance has low receptivity at natural abundance due to the low γ. An extensive chemical shift range makes the signal highly susceptible to environmental parameters such as the tem-perature gradient, broadening the signal. The linewidths of nuclei with a spin quantum number of $\frac{1}{2}$ are generally small, so the signals are sharp unless affected by paramagnetism. Nuclei with a spin quantum number $>\frac{1}{2}$ have inherent electric quadrupole moment that broadens the signal.

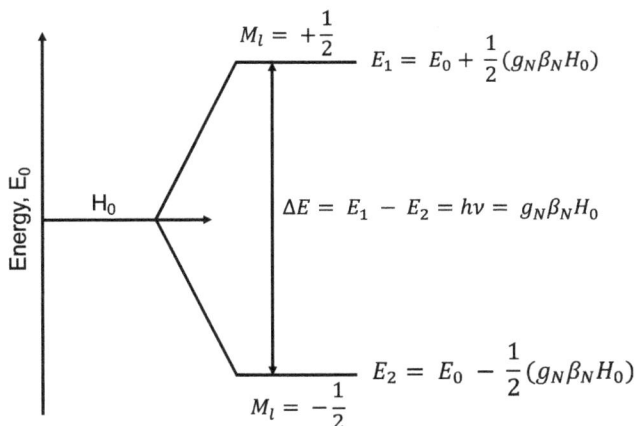

Figure 4.10 The basic principle of NMR of a proton with a spin quantum number, M_I, of $\frac{1}{2}$ placed in an external magnetic field, H_0. The energy diagram shows the Zeeman splitting, whereby the precessing proton in the lower energy state, $M_I = -\frac{1}{2}$, absorbs energy, ΔE, from a radiofrequency source to pass into the higher energy state ($M_I = +\frac{1}{2}$). G_N is the nuclear g-factor, which is 5.5855 for a proton and β_N is a nuclear magneton, which is 5.0509×10^{-27} A m^2.

nuclear spin quantum number $\geq \frac{1}{2}$, a property that enables a charged nucleus to spin about an axis, generating a magnetic dipole moment. In an external magnetic field, this property makes NMR active nuclei align with or against the field, enabling the absorption of radiofrequency radiation, the foundation of NMR spectroscopy (Figure 4.10).

In addition, for nuclei with a spin quantum number greater than $\frac{1}{2}$, the charge distribution of the protons is asymmetric, producing an electric quadrupole moment alongside the magnetic dipole moment. Nuclei having quadrupole moments constitute over 66% of all NMR active nuclei. The presence of quadrupole moment shortens spin–lattice relaxation time – the time it takes for an excited magnetisation vector to return to equilibrium – and significantly broadens the NMR signals, precluding the observation of well-resolved signals. For example, ^{59}Co is an NMR active nucleus with a natural abundance of 100% and a spin of $\frac{7}{2}$, which confers an electric quadrupole moment. Together, these properties yield highly sensitive but broad signals in a ^{59}Co NMR experiment. Signal broadening also occurs when nuclei coupled to others with electric quadrupole moment.

A second criterion, which is just beneficial but unnecessary, is that the nucleus occurs in significant natural abundance. Relative to the proton with natural abundance $> 99.9\%$, tin-119, ^{119}Sn, has a natural abundance of only 8.58% but is still used to characterise tin-containing polymers.[15] In an extreme situation where low natural abundance significantly lowers the signal-to-noise ratio, precluding accurate NMR spectroscopic investigation, isotope enrichment is a strategy to bolster sensitivity.

Another criterion is that the nucleus possesses a relatively short spin–lattice relaxation time (T_1); however, the deleterious effect of a long T_1 can be ameliorated by using pre-acquisition delay times. Another consideration for organometallic and coordination polymers is paramagnetic centres, which have one or more unpaired electrons. The presence of paramagnetic centres significantly affects NMR signals; specifically, they widen the chemical shift range and broaden the signal. Broadening due to the presence of paramagnetic species can be so great as to distort signals. As the various oxidation states of transition metals can exist as diamagnetic or paramagnetic species, it is still possible to observe an NMR signal by choosing a diamagnetic oxidation state. For example, ^{59}Co NMR is not observable for Co^{2+}, the most common oxidation state in cobalt compounds, because it is paramagnetic but observable for Co^{3+} compounds that are diamagnetic.

Provided the signal is unaltered by paramagnetism and anisotropy, proton NMR sensitivity is enough to characterise the composition and structure of inorganic and organometallic polymers.[16] For example, ^{1}H NMR can be used to confirm an organometallic moiety in a polymer. As a specific example, the η^6-*p*-disubstituted benzene–η^5-cyclopentadienyliron(II) complex has a distinctive, well-resolved ^{1}H NMR peak in the spectrum of various polymers, making it possible to confirm its presence in these polymers. In this complex, the iron is sufficiently diamagnetic, allowing the observation of the proton peaks of the disubstituted benzene and the cyclopentadienyl ligands (Figure 4.11). The resonance frequencies of the protons in the ^{1}H NMR spectrum of the η^6-*p*-disubstituted benzene–η^5-cyclopentadienyliron(II)-based dendrimer can also be used to determine the dendrimer generation. Other nuclei such as phosphorus and boron, although less sensitive than a proton, can be used to characterise the composition of a polymer network. In the polycationic organometallic dendrimer in Figure 4.11, the presence of a phosphorus hexa-fluoride or boron tetrafluoride counteranion was confirmed using ^{31}P or ^{11}B NMR, respectively.[14]

When a stereogenic centre is present, ^{1}H and ^{13}C NMR are valuable tools to ascertain the polymer tacticity and determine the statistical distribution of the various stereogenic centres. *In situ* ^{1}H NMR also allows monitoring of polymerisation reactions and various dynamics in the polymer system. Again, ^{1}H and ^{13}C NMR are among the routine techniques used to evaluate the composition and distribution of monomers in a co-polymer.[17] The peak integrals can be used to estimate the ratio of the monomers in the copolymer. Last, since NMR linewidth is affected by the mobility of the molecules, it is possible to estimate the microscale ordering of polymer chains, precisely, the degree of crystallinity of semi-crystalline polymers using these methods. The rationale for determining the degree of crystallinity is that, above the glass transition temperature (T_g), the polymer chains in the amorphous domain perform detectable segmental motions within the NMR timescale, producing sharp peaks. In contrast, chains in the crystalline domains are less mobile, giving rise to broad peaks. The degree of crystallinity is defined by eqn (4.34).

$$\text{Degree of crystallinity} = \frac{\text{Integral of broad peak}}{\text{Integral of broad peak} + \text{Integral of sharp peak}} \qquad (4.34)$$

4.3.2 Electron Paramagnetic Resonance Spectroscopy

Electron paramagnetic resonance (EPR) spectroscopy, also known as electron spin resonance spectroscopy, is used to study paramagnetic species with one or more unpaired electrons. Therefore, the method can be used to determine the presence of a

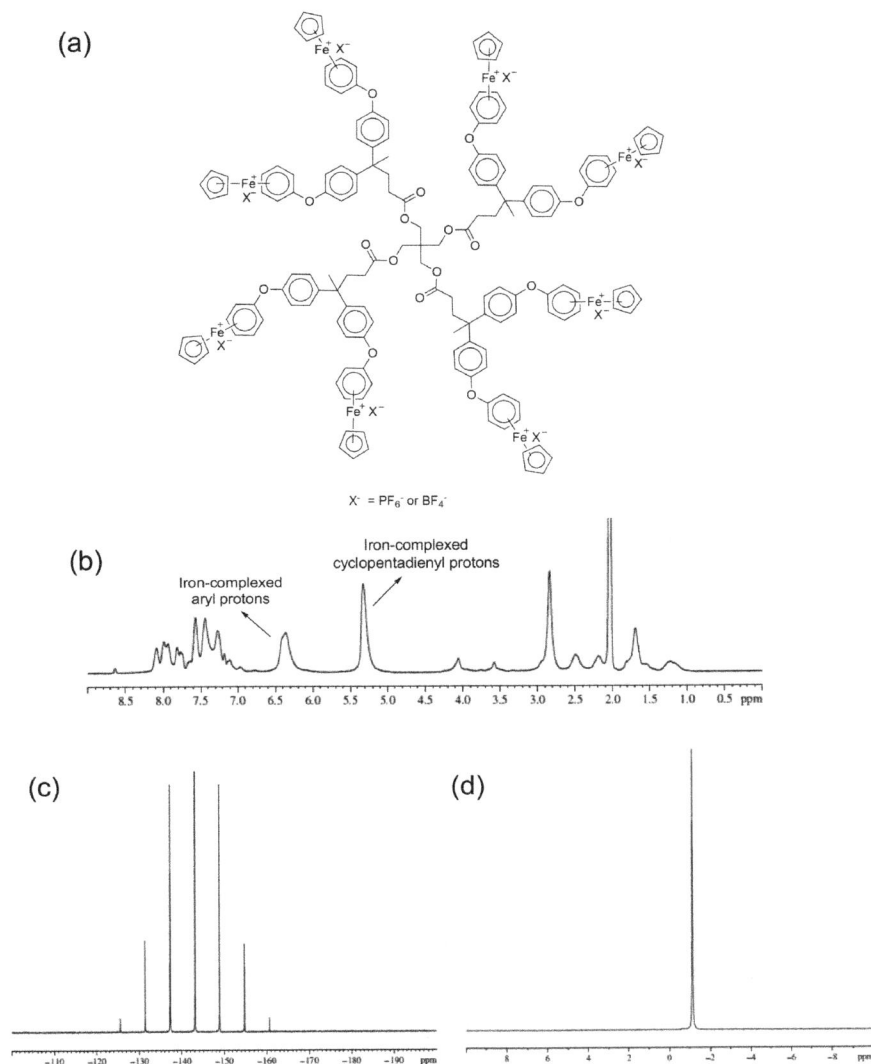

Figure 4.11 (a) An organometallic dendrimer characterised by ^{1}H NMR, ^{31}P NMR, and ^{11}B NMR spectroscopy. (b) The ^{1}H NMR spectrum shows a peak with resonance frequencies corresponding to the protons of the iron-complexed cyclopentadienyl and aryl ligands in the dendrimer, confirming the organometallic identity of the dendrimer. (c) The ^{31}P NMR spectrum shows peaks with a resonance frequency corresponding to phosphorus in the PF_6 counteranion. (d) The ^{11}B NMR spectrum shows peaks with a resonance frequency corresponding to boron in the BF_4 counteranion. (b) Adapted from ref. 21 with permission from John Wiley and Sons, Copyright 2014 Wiley-VCH Verlag GmbH & Co. KgaA, Weinheim. (c and d) Reproduced from ref. 14 with permission from American Chemical Society, Copyright 2015.

paramagnetic species in inorganic and organometallic polymers. Polymers containing paramagnetic metal ions such as Fe^{3+} (d^5), V^{4+} (d^1) or Co^{2+} (d^7) with unpaired electrons will exhibit an EPR spectrum. Like other spectroscopic methods, EPR spectroscopy involves the absorption of electromagnetic radiation, in this case, microwave radiation, to induce a transition between the magnetic energy levels of electrons that

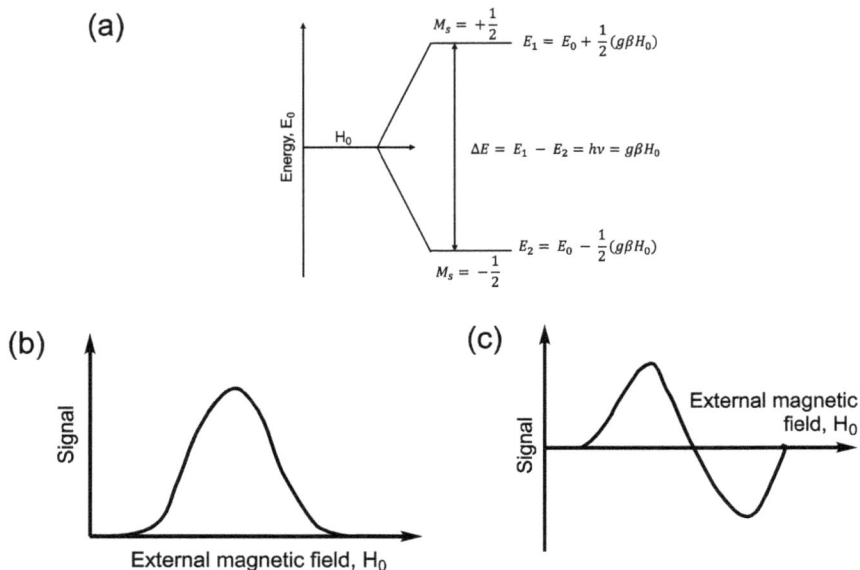

Figure 4.12 (a) The basic principle of EPR spectroscopy depicted with an energy diagram showing Zeeman splitting of an electron after absorption of energy ΔE from a microwave frequency source. G is the Landé g-factor (g-value). (b) A typical EPR absorption spectrum and (c) first derivative form of the absorption spectrum.

possess spin and orbital angular momenta. Upon absorption of appropriate microwave radiation (ΔE) by the sample, a transition between two spin states occurs, resulting in an absorption spectrum, which is transformed to a more sensitive first derivative form by a phase-sensitive detector (Figure 4.12).

The Landé g-factor (g-value) for the sample is determined from the magnetic field, B_{sample}, that gives the absorption maximum corresponding to the zero point on the derivative curve (Figure 4.12). The experimental g-value, obtained by substituting B_{sample} for B_0 in eqn (4.33), provides information on the number of unpaired electrons, the coordination sphere, and the molecular symmetry. For example, EPR of an Fe^+-containing dendrimer gives three g-values to support the rhombic distortion of η^6-arene–η^5-cyclopentadienyliron(I) complexes.[18] Also, the method can be used to study polymer dynamics in solution, as exemplified by the observed redox behaviour of the Fe^{2+}-containing dendrimers in solution (Figure 4.13).[19]

$$\Delta E = h\nu = g\mu_B B_0 \tag{4.35}$$

g is the Landé g-factor, μ_B is the Bohr magneton and is equal to 9.2740×10^{-24} JT^{-1}, and B_0 is the applied magnetic field.

4.3.3 Electronic Spectroscopy

Electronic spectroscopy entails measuring the electromagnetic radiation associated with the electronic transition between energy levels. Transitions from low to high energy levels produce an absorption spectrum, while those from high to low energy levels

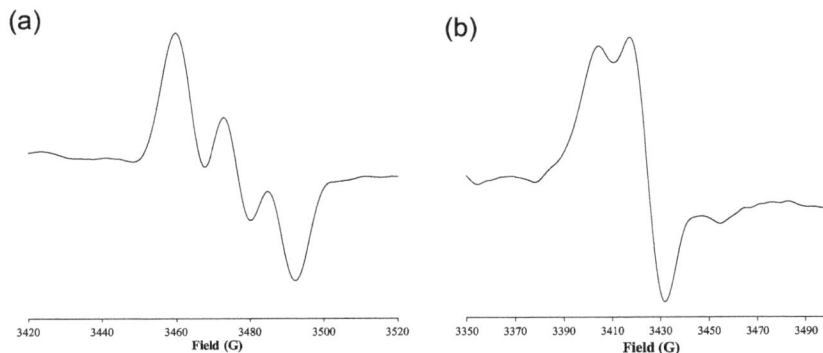

Figure 4.13 Electron paramagnetic resonance unravels the organometallic dendrimer reaction mechanism, confirming that iron-containing dendrimers in solutions generate free radicals. EPR spectra of a radical generated by reacting a dendrimer with a spin trap, nitrosobenzene, measured at (a) room temperature and (b) 77 K. At room temperature, the radical gives an isotropic EPR resonance at $g_{iso} = 2.008$ and at 77 K an anisotropic nitrogen-split triplet EPR resonance at $g_{1,2,3} = 2.010$, 2.005, and 1.998. Reproduced from ref. 19 with permission from John Wiley and Sons, Copyright 2017 Wiley-VCH Verlag GmbH & Co. KgaA, Weinheim.

give an emission spectrum. Therefore, electronic spectroscopy comprises two types: absorption and emission spectroscopy. This section will focus on how absorption spectroscopy elucidates the molecular structure, and Section 4.4 will discuss how emission spectroscopy, which includes fluorescence and phosphorescence, characterises bulk macroscopic properties.

Both atomic and molecular species produce electronic spectra, but we will discuss molecular species, which are more relevant to polymers. Electronic spectroscopy involves electronic, vibrational, and rotational states and therefore captures a snapshot of a molecule in a particular state. We must remember that the energy absorbed during an electronic transition from the ground, low energy, state to the excited, high energy, state corresponds to the ultraviolet (UV) or visible region of the electromagnetic spectrum.

The absorption spectra of coordination and organometallic polymers arise from transitions involving $\pi^* \leftarrow \pi$, $\sigma^* \leftarrow n$, $\pi^* \leftarrow n$, d–d, or metal-to-ligand or ligand-to-metal charge transfer. In most cases, the $\pi^* \leftarrow \pi$, $\sigma^* \leftarrow n$, and $\pi^* \leftarrow n$ transitions result from the organic moiety, the d–d transition from the metal centre, and the charge transfer transition from electronic transfer between metal and ligand orbitals. For example, the UV-vis absorption spectrum of a coordination polymer, $\{[Co_2(L)(biim-4)]\cdot 3H_2O\}_n$ where biim-4 is 1,1′-(1,4-butanediyl)bis(imidazole) and L is 1,4,8,11-tetrazacyclotetradecane-*N,N,N,N*-tetramethylene-benzoate, consists of absorption bands at 200–350 nm, 576 nm, 797 nm, and 1096 nm.[20] The band at 200–350 nm results from $\pi^* \leftarrow \pi$, $\sigma^* \leftarrow n$, and $\pi^* \leftarrow n$ transitions in the organic moieties of the polymer, while the bands at 576 nm, 797 nm, and 1096 nm result from d–d transitions (Figure 4.14). The d–d transition, except in a gas-phase ion, has weak intensity because the Laporte selection rule forbids it, but the vibronic coupling in these polymers enables this transition, leading to a strong absorption band (Figure 4.14).

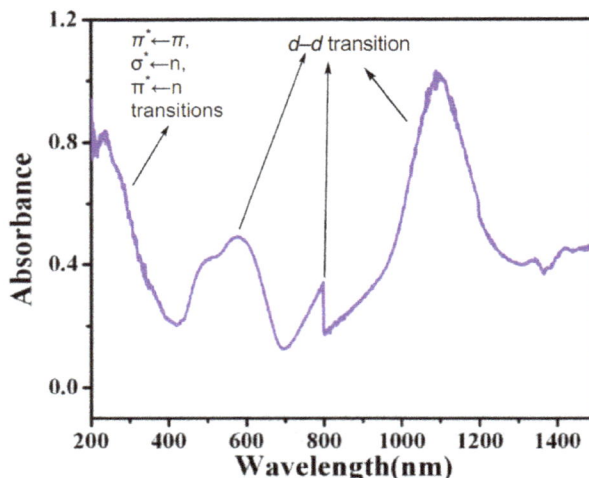

Figure 4.14   UV-vis absorption spectrum of a coordination polymer showing bands attributed to $\pi^* \leftarrow \pi$, $\sigma^* \leftarrow n$, $\pi^* \leftarrow n$, and d–d transitions. The band at 200–350 nm is attributed to $\pi^* \leftarrow \pi$, $\sigma^* \leftarrow n$, and $\pi^* \leftarrow n$ transitions in the organic moieties of the polymer, while the bands at 576 nm, 797 nm, and 1096 nm result from d–d transitions. Reproduced from ref. 20 with permission from Elsevier, Copyright 2016.

The absorption band due to charge transfer transitions is commonly observed in the UV-vis spectra of coordination and organometallic polymers. Bands from charge transfer transitions may be observed as absorption tails at a longer wavelength. For example, the UV-vis absorption spectra of an organometallic dendrimer show bands at short wavelengths due to the typical $\pi^* \leftarrow \pi$, $\sigma^* \leftarrow n$, and $\pi^* \leftarrow n$ transitions, and another band attributed to the charge transfer transition (Figure 4.15).[21] The intensity of the charge-transfer absorption band increases with increasing dendrimer generation and increased electronic interactions between ligand and metal orbitals. In summary, the electronic spectra of coordination and organometallic polymers allow us to understand the extent of conjugation in a polymer, the degree of interaction between ligands and the metal centre, the oxidation states of the metals, and the electronic communication between different segments of a polymer.

4.3.4   Infrared and Raman Spectroscopy

Infrared (IR) and Raman spectroscopy involve transitions between vibrational energy levels, with both methods being complementary. A vibration is IR active if it induces a change in dipole moment and Raman active if it causes a change in polarisability. An IR spectrum can be divided into two regions: a region of bands above 1500 cm^{-1}, which provides information on specific functional groups, and a region below 1500 cm^{-1}, which provides a diagnostic fingerprint of a particular molecule. The ability of IR spectroscopy to identify absorption due to a specific functional group or bond makes it a valuable tool to monitor polymerisation progress and completion. This involves monitoring the disappearance or appearance of functional groups in the monomer or polymer, respectively. For example, IR spectroscopy monitors the complete synthesis of polyphosphazene-*co*-polysiloxane copolymers by monitoring the disappearance of the

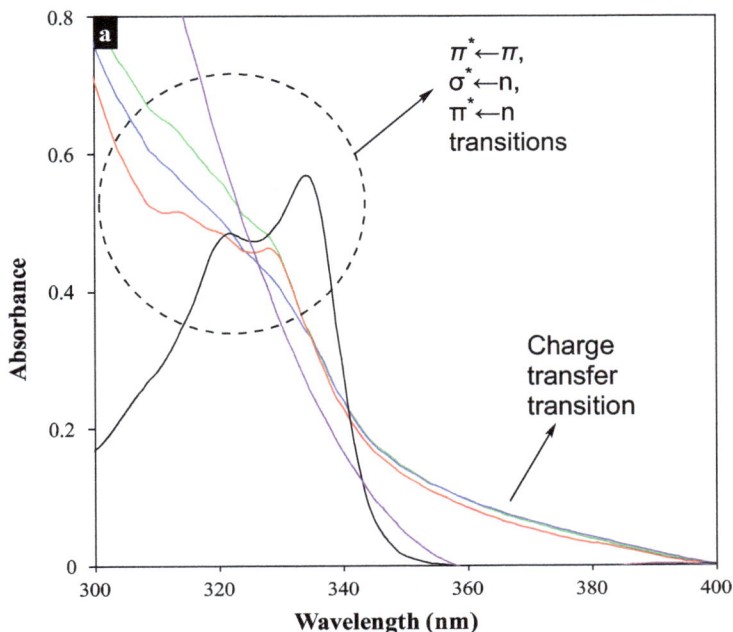

Figure 4.15 UV-vis absorption spectra of β-naphthol-capped organometallic dendrimers show-
ing bands attributed to $\pi^* \leftarrow \pi$, $\sigma^* \leftarrow n$, $\pi^* \leftarrow n$, and charge transfer transitions. The
band at 280–350 nm is attributed to $\pi^* \leftarrow \pi$, $\sigma^* \leftarrow n$, and $\pi^* \leftarrow n$ transitions in the
organic moieties of the dendrimers, while the absorption tail above 350 nm results
from charge transfer transitions. Black: β-naphthol, purple: 1st generation dendron,
green: 1st generation dendrimer, blue: 2nd generation dendrimer, and green: 3rd
generation dendrimer. Reproduced from ref. 21 with permission from John Wiley and
Sons, Copyright 2014 Wiley-VCH Verlag GmbH & Co. KgaA, Weinheim.

Si–H stretching mode, which absorbs at 2100 cm^{-1}, in the poly(dimethylsiloxane)
macromonomer.[22] In another example, the appearance of carbonyl stretching
bands at 2095, 2057, and 2024 cm^{-1} in the IR spectra indicates the successful coordination of cobalt hexacarbonyl to alkyne-containing organometallic dendrimers to
form heterobimetallic dendrimers.[23] Qualitative application of IR spectroscopy to
polymer research includes structural elucidation by identifying the functional
groups present in the polymer and confirming the monomer sequence distribution in
copolymers.

Raman spectroscopy is more valuable than IR spectroscopy for characterising molecules containing homonuclear bonds that exhibit minimal or no change in dipole
moment upon vibrational transitions. For example, the symmetrical sulphur–sulphur
bond in polysulphur or the silicon–silicon bond in polysilane undergoes stronger
stretching, bending, and deformation vibrations in Raman spectroscopy than in IR.
Therefore, Raman spectroscopy enables detailed characterisation of inorganic polymers
containing symmetrical bonds. A typical example is the Raman spectroscopy characterisation of a sulphur–selenium copolymer, where the spectrum provides information
on the formation of sulphur–sulphur, sulphur–selenium, and selenium–selenium
bonds. The Raman spectrum of the polysulphur contains three peaks at 467, 220, and
150 cm^{-1} corresponding to the stretching, bending, and deformation vibrations of the

sulphur–sulphur bond, respectively.[24] On incorporating selenium to form the sulphur–selenium copolymer, new peaks corresponding to sulphur–selenium and selenium–selenium stretching modes appear at 375 and 194 cm^{-1}, respectively. The intensity of these peaks increases with an increase in the content of selenium in the polymer. Compared to IR spectroscopy, Raman spectroscopy is less popular as a characterisation tool for inorganic and organometallic polymers due to the low intensity of the Raman scattering, but this can be circumvented using Fourier transform techniques. Nevertheless, an advantage of Raman spectroscopy over IR spectroscopy is the possibility to extend into low wavenumber regions, permitting the study of vibrational modes that are non-observable in IR spectroscopy.

4.3.5 X-ray Diffraction Methods

Electrons surrounding the atomic nuclei diffract or scatter X-rays, enabling structural characterisation of molecular and non-molecular solids. The X-ray diffraction method is based on Bragg's equation (eqn (4.36)) which relates the wavelength, λ, of the incident X-ray radiation to the lattice spacing, d, of the crystal and the grazing angle, θ (Figure 4.16). The degree of diffraction depends on the electron density, providing a framework to differentiate atoms. Usually, the diffraction data collected over a range of θ values and crystal orientations generate a three-dimensional picture of electron density. The picture contains information on the type of atom, the atomic position in the crystal, and mean chemical bond length and angle.

$$2d \sin \theta = n\lambda \qquad (4.36)$$

Different X-ray diffraction techniques exist to characterise materials, with the most widely used being single-crystal X-ray diffraction, powder X-ray diffraction, and small-angle X-ray scattering. A single crystal, usually grown from a solution, scatters a beam of irradiated X-ray to produce a well-defined diffraction pattern of spots. Provided the crystal is of sufficient purity and regularity, the diffraction data can be used to determine the mean chemical bond length and angle. Single crystal diffraction analysis has

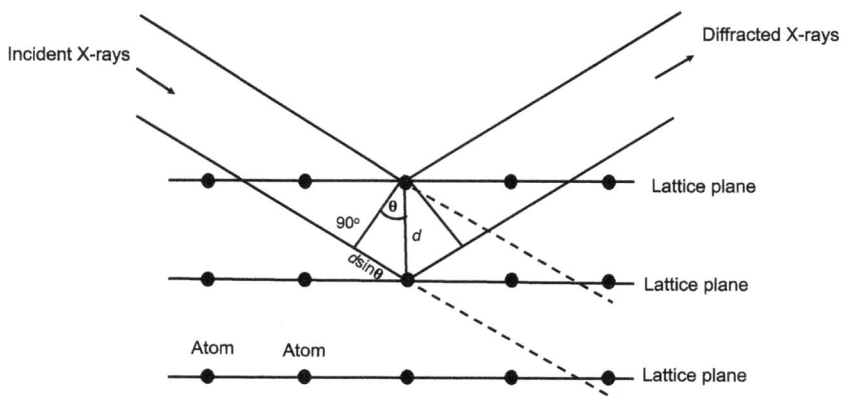

Figure 4.16   The geometry of diffraction of X-rays by a layer of atoms in a crystal.

been used to determine the dimensionality, bond angle and length, and space group of coordination polymers. For example, single-crystal diffraction shows that some co-ordination polymers are one-dimensional, linear polymers belonging to the $P2_1/n$ space group.[25] This diffraction method also reveals the role of supramolecular interactions such as hydrogen bonding, π–π and CH–π interactions in the structural organisation of coordination polymers.

For polycrystalline materials, which include some coordination polymers, that are incapable of forming single crystals, the powder X-ray diffraction (pXRD) technique can be used under certain conditions to solve the molecular structure.[26] In pXRD, because the microcrystals lie in random orientation, the X-ray interacts with the sample and scatters in all directions. In contrast with the single crystal technique where the output is a pattern of sharp spot, in pXRD, the pattern consists of concentric rings but with the same scattering angle, 2θ, as the spot in a crystal pattern. This technique helps characterise different phases in ceramics derived from organometallic polymers and, in some cases, the degree of crystallinity in semi-crystalline polymers. Most polymers are semi-crystalline or amorphous; therefore, the most widely used X-ray technique for polymer characterisation is small-angle X-ray scattering (SAXS). SAXS characterises the nanoscale structural and morphological properties of polymers; for example, it provides morphological information on the size and shape of nanostructures such as lamellae, spherulite, and micelles.

4.3.6 Differential Scanning Calorimetry

Because we mentioned the degree of crystallinity, a morphological property, we should consider differential scanning calorimetry (DSC), a thermoanalytical technique used to estimate the same property. Data obtained from a DSC experiment can be used to determine the degree of crystallinity in bulk polymer samples. A DSC experiment measures the amount of heat flowing in and out of a sample as a function of time or temperature, data that can be used to compute melting (T_m), crystallisation (T_c), and glass transition (T_g) temperatures (Section 4.4). The enthalpy of transition (ΔH) is determined by integrating the peak area (A) corresponding to that transition, and both quantities are related by eqn (4.37).

$$\Delta H = KA \tag{4.37}$$

K is the calorimetry constant, specific to a DSC instrument and determined using a standard with a known enthalpy of transition. The percentage crystallinity is estimated using eqn (4.38).

$$\% \ \text{Crystallinity} = \left(\frac{\Delta H_m - \Delta H_c}{\Delta H_{m^c}} \right) \times 100 \tag{4.38}$$

ΔH_m and ΔH_c are the enthalpies associated with melting and crystallisation transitions, while ΔH_{m^c} is the enthalpy of melting for the 100% crystalline sample. The ΔH_{m^c} values are established for common polymers and found in the literature.[27] Depending on the thermal history of the sample, a T_c may be absent; in that case, the percentage crystallinity

Box 4.7 Worked example 4.5.

Question

Using the data below, calculate the percentage crystallinity of the following inorganic polymers assuming the absence of a crystallisation exotherm in the DSC thermogram. Comment on the effect of changing the alkyl group on the degree of crystallinity.

Polymer	ΔH_m (J g^{-1})	ΔH_{mc} (J g^{-1})
Poly(dimethyl siloxane)	10.0	52
Poly(diethyl siloxane)	10.5	55

Answer

Recall eqn (4.39)

Poly(dimethyl siloxane)

$$\text{Percentage crystallinity} = \left(\frac{10 \text{ Jg}^{-1}}{52 \text{ Jg}^{-1}}\right) \times 100 = 19.2\%$$

Poly(diethyl siloxane)

$$\text{Percentage crystallinity} = \left(\frac{10.5 \text{ Jg}^{-1}}{55.0 \text{ Jg}^{-1}}\right) \times 100 = 19.1\%$$

Changing from the methyl to ethyl group does not affect the degree of crystallinity.

is obtained using eqn (4.39). Modern DSC instruments are equipped with computer programs that compute the percentage crystallinity (Box 4.7).

$$\% \text{ Crystallinity} = \left(\frac{\Delta H_m}{\Delta H_{mc}}\right) \times 100 \tag{4.39}$$

4.3.7 Microscopy

Microscopy involves studying and using microscopes to examine the morphology and topology of materials. Data generated from microscopy are spatially resolved into images, enabling visualisation of materials at high magnification. Most microscopic methods irradiate the investigated materials with an energy source – photons, electrons, or ions – which can stimulate the materials to emit electrons, photons, or X-rays. A suitable detector can analyse these low-energy emissions to provide spectroscopic or crystallographic data to characterise the chemical composition or structure of the material. Some modern microscopes are powerful instruments that produce high-resolution images and spectroscopic and crystallographic data to characterise the morphology and topology of materials. Various microscopic methods exist, but we will discuss scanning electron microscopy (SEM) and transmission electron microscopy (TEM) which are widely used to characterise inorganic and organometallic polymers.

A thermionically generated electron beam of 0.2 keV to 40 keV energy is scanned across a sample surface in scanning electron microscopy. The electron beam interacts with the sample, leading to the emission of a low-energy secondary electron by inelastic scattering, emission of electromagnetic radiation, absorption of high-energy primary electrons, and back-scattering of the high-energy primary electrons by elastic scattering, each of which can be detected and used to produce an image with a well-defined, three-dimensional appearance. Detection of the emitted photons has analytical capabilities, for example, in the elemental analysis of the sample surface as in energy-dispersive X-ray spectroscopy or analysis of the intensity of electron-induced luminescence as in cathodoluminescence microscopy.

SEM analysis of polymers faces two key challenges. First, because the image contrast from electron microscopes results from electron scattering by atomic nuclei, polymer samples consisting of carbon, hydrogen, oxygen, and nitrogen produce poor contrast. However, organometallic and coordination polymers containing transition metals are likely to generate images with improved contrast. Also, most polymers are poor conductors of electricity; therefore, the generated secondary electrons accumulate on the sample surface to interact with the incident primary electrons, leading to distorted images. This electron build-up may not be a significant problem for coordination and organometallic polymers that contain metals, but it is advisable to coat the sample surface with a thin film of conducting materials such as gold or palladium to earth the build-up charges to obtain stable images. Typical sample preparation for organometallic polymers involves sprinkling a sample on a double-sided copper tape adhered on aluminum stubs, then sputter-coating with palladium for 10 seconds before imaging. This sample preparation permits the use of SEM to obtain high-resolution images of the surface topology and composition of the organometallic dendrimers discussed in Figure 4.11a (Figure 4.17a).[23] Although SEM has limited application in studying surface morphology, it can provide helpful information on surface topology and composition.

Transmission electron microscopy (TEM) uses transmitted electrons, electrons that pass through the sample, to acquire two-dimensional images (this contrasts with SEM, where back-scattered or emitted secondary electrons are used to generate three-dimensional images). As a result, TEM provides vital information on morphology and structures. Because TEM detects transmitted electrons, samples must be very thin, usually less than 100 nm, a requirement that makes sample preparation more tedious than in SEM. For organometallic polymers, the sample can be dispersed in a solvent,

Figure 4.17 Microscopic examination of an organometallic dendrimer: (a) scanning electron microscopy and (b) transmission electron microscopy of the dendrimer. Reproduced from ref. 23 with permission from the Royal Society of Chemistry.

which does not solubilise the polymers, pipetted onto a carbon-coated copper grid, then air-dried before imaging. This sample preparation allows high-resolution imaging of the organometallic dendrimers in Figure 4.11a (Figure 4.17b).[23] Other sample preparation techniques include freeze-drying of the polymer solution. The spatial resolution of TEM is superior to that of SEM, with the latter currently limited to ~0.5 nanometres while the former approached 50 picometres. What decides the choice of a microscopic technique depends on the information the polymer chemist seeks. It is best to use SEM if the surface topology needs to be investigated and TEM if the inner morphology is the study's goal.

4.4   Characterisation of Macroscopic Properties

Macroscopic properties define the use of materials. It is of the utmost importance to fully characterise these properties to understand the potential application of polymeric materials. For a polymer, temperature-dependent properties, such as thermal stability and glass transition and melting temperatures, and time-related phenomena such as creep or stress relaxation need to be ascertained to understand and predict how temperature or time affects the performance of these materials during use. We will discuss analytical methods, such as rheology and thermal analysis, that provide this information. Also, we will discuss methods used to characterise functional properties, exemplified by photoactivity, redox activity, conductivity, magnetic properties, and bioactivity. Indeed, introducing functional properties into polymers is one reason polymer chemists design and synthesise inorganic and organometallic polymers.

4.4.1   Thermal Analysis

Undoubtedly, thermal analysis is one of the essential characterisations carried out on polymeric materials intended for everyday use. Therefore, most polymer laboratories are equipped with a thermogravimetric analyser to determine the thermal stability and a differential scanning calorimeter to evaluate crystallisation, melting and glass transition temperatures. We introduced differential scanning calorimetry in Section 4.3.5 as a tool to determine the degree of crystallinity. In this section, we will present it as a method to quantitatively evaluate the crystallisation temperature (T_c), melting temperature (T_m) and glass transition temperature (T_g) of polymers. To recap, DSC is a thermoanalytical method whereby a polymer sample is simultaneously heated with a reference under an inert atmosphere, and the thermal transitions in the sample are detected (Figure 4.18). The underlying principle of DSC is that the heat flow to the polymer sample during a thermal transition differs from that to the reference, provided both are maintained at the same temperature. The differential power ($d\Delta Q/dt$) needed to maintain the sample and reference at the same temperature is recorded as a function of temperature.

A plot of $d\Delta Q/dt$ (heat capacity) against temperature yields a thermogram (Figure 4.18), which contains peaks corresponding to T_g, T_c, and T_m. After calibration, the peak area is directly related to the enthalpy change in the sample and may estimate thermodynamic parameters such as enthalpies of reaction. Let us consider a typical

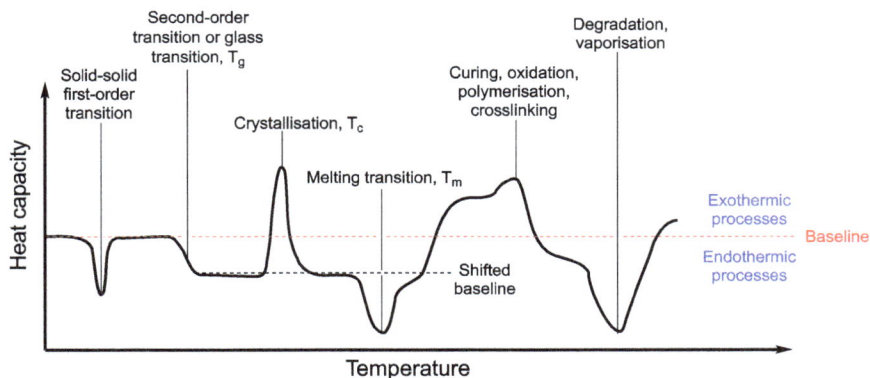

Figure 4.18 A hypothetical DSC thermogram shows expected transitions in a semi-crystalline polymer. The first-order transitions are physical transitions from structures of different sizes and shapes in the polymer. Chemical reactions such as polymerisation and curing may be endothermic or exothermic, while oxidation and cross-linking are always endothermic.

DSC characterisation of a hypothetical semi-crystalline polymer consisting of amorphous and crystalline domains. As the programmed temperature ramps up, the amorphous domains transition from a rigid "glassy" state to a viscous or "rubbery" state. This transition is an endothermic process, resulting in a change in the heat capacity of the sample without a formal phase change that shifts the initial baseline of the recorded thermogram, corresponding to T_g. With further increase in temperature, the amorphous domain becomes less viscous, gaining sufficient freedom of motion to realign into crystalline domains. The transition from an amorphous to a crystalline domain is an exothermic process that also produces a peak corresponding to T_c. As the temperature increases, the crystalline domains begin to melt, an endothermic process that yields a peak corresponding to T_m. Whereas amorphous polymers such as a ruthenium-containing coordination polymer exhibit only T_g,[28] semi-crystalline polymers are more likely to show all transitions. Conventionally, the onset of the transition, the inflexion point in the peak or the peak maximum can be used to depict the transition temperature. It is also important to note that DSC can be used to study the polymerisability of organometallic monomers and the enthalpy of polymerisation, which appear as exothermic or endothermic peaks in the thermogram.

Another thermal analysis is thermogravimetric analysis (TGA), used to determine the thermal stability and oxidative stability of polymers. This analysis involves measuring the weight change of a sample subjected to programmed heating in an air- or inert gas-filled furnace as a function of time or temperature. The most widely used TGA method, non-isothermal TGA, involves continuously measuring the weight change of the sample as the temperature is ramped up at a controlled heating rate. Data are presented as a thermogram, a plot of weight change against temperature or time (Figure 4.19a). The data can also be expressed as a first derivative of the TGA signal to precisely define the onset and endset of subtle weight change events that characterise the sample (Figure 4.19b).[29] Weight changes typically result from the loss of residual solvents, oxidation, or polymer decomposition and correspond to a slope or a shift in the baseline of the thermogram (Figure 4.19a). Thermally stable polymers

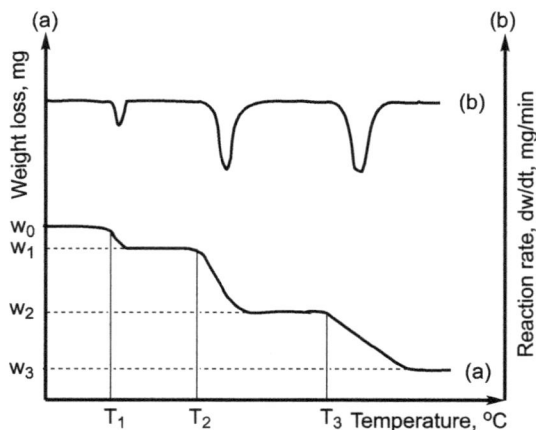

Figure 4.19 (a) A TGA thermogram showing typical weight changes (w_1, w_2, w_3) and decomposition temperature (T_1, T_2, and T_3). (b) The first derivative of the TGA (DTG) curve used for in-depth analysis, such as kinetic studies.

exhibit no or little weight change that corresponds to a lack of slope or shift of the initial baseline of the TGA trace within a time or temperature range. To evaluate oxidative stability, the TGA experiment is conducted in an air-filled furnace instead of an inert gas-filled furnace with oxidation resulting in a weight gain. In an isothermal TGA, where the weight change is monitored as a function of time at a constant temperature, oxidative stability is the time between introducing air into the furnace and the onset of weight gain. Although the primary use of TGA is thermal and oxidative stability studies, the technique also provides information on the decomposition kinetics and mechanism of polymers. Modern TGA instruments can be coupled to a mass spectrometer or an IR spectrometer to detect and characterise effluent decomposition products.

4.4.2 Dynamic Mechanical Analysis

Dynamic mechanical analysis (DMA) assesses molecular mobility in polymers. DMA data can be used to delineate the viscoelastic properties of polymers and identify various motional transitions in the glassy or crystalline state, such as T_g. A typical dynamic mechanical analyser applies a small sinusoidally varying stress on the test polymer and measures the strain to compute complex modulus. The resulting strain is in phase with the applied stress if the sample is an entirely elastic solid but out of phase by 90° if it is a purely viscous Newtonian fluid. Depending on the utilisation temperature, polymers are largely viscoelastic materials, exhibiting both elastic and viscous behaviours, and therefore show some phase lag between strain and stress during DMA experiments. Usually, DMA data are resolved into an in-phase elastic-like component, represented by the storage modulus, E' or G', that measures the stored energy, and an out-of-phase viscous-like component, represented by the loss modulus, E'' or G'', that measures the dissipated energy. The storage modulus and loss modulus can be used to determine T_g, polymer composition, and miscibility. At T_g, for instance, the storage modulus decreases dramatically, while the loss modulus increases to a maximum.

4.4.3 Electronic Spectroscopy

Electronic spectroscopy, as absorption spectroscopy or photoluminescence, is a valuable method to characterise the optical and electronic properties of materials. Section 4.3.3 discusses how absorption spectroscopy characterises the molecular structure and dynamics, but here we will examine how the technique and photoluminescence probe the optical and electronic properties. From the chemistry perspective, photoluminescence refers to the spontaneous emission of photons from a photoexcited species and includes both fluorescence and phosphorescence. Photoexcitation usually occurs from a singlet ground state to a singlet first excited state, which can relax back to the ground state or a low energy excited triplet state. In fluorescence, the decay occurs without a change in multiplicity, with the excited species undergoing a singlet-to-singlet electronic relaxation to the ground state, leading to the emission of a photon. In contrast, phosphorescence involves intersystem crossing from the singlet first excited state to the lowest energy triplet state, followed by a triplet-to-singlet electronic relaxation through the emission of a photon. The time between excitation and emission differs for both phenomena, with fluorescence having lifetimes in the picosecond to nanosecond regime and photoluminescence being a spin-forbidden transition, requiring a millisecond to decasecond lifetime.

A prerequisite for photoluminescence is the presence of a species amenable to photoexcitation and decay of the excited states back to the ground state with the emission of a photon (Figure 4.20). The presence of these species in an inorganic and organometallic polymer can be characterised using steady-state or time-resolved photoluminescence instrumentations. Usually, the information sought from photoluminescence techniques includes luminescence intensity, excitation and emission spectra, lifetime, and polarisation. The information allows characterisation of the electronic and optical properties and a structural disorder in the polymer. For example, lifetime data not only distinguish fluorescent from phosphorescent polymers but also elucidate the structure–property relationship. In an organometallic dendrimer, changes in lifetimes between dendrimer generations indicate backfolding of peripheral structures into the dendrimer core with increasing generation. Lifetime data can also confirm electronic interactions within the polymer structure through observation of the photoluminescence decay process.[21] Structural disorder or the presence of impurities in the polymer provides pathways for the decay of the excited states to the ground states with a monoexponential or biexponential lifetime decay involving the presence of one or two different pathways, respectively.[30]

The optical band gap of organometallic polymer-based semiconductors is a critical property required to determine electrical conductivity and select materials for photovoltaic design. For organometallic pi-conjugated polymers, the bandgap corresponds to the energy difference between the highest occupied molecular orbital and the lowest unoccupied molecular orbital, analogous to the valence and conduction bands of inorganic semiconductors, respectively. The optical bandgap is related to the electronic band structure of pi-conjugated systems such as platinum(II) polyyne and is experimentally obtained from electronic spectroscopy, specifically UV-vis absorption spectroscopy, of the polymer solution or thin film. The optical bandgap (E_g^{opt}) can be estimated from the absorption edge wavelength ($\lambda_{a.e}$) using eqn (4.40), where h is Planck's constant and c is the velocity of light.[32]

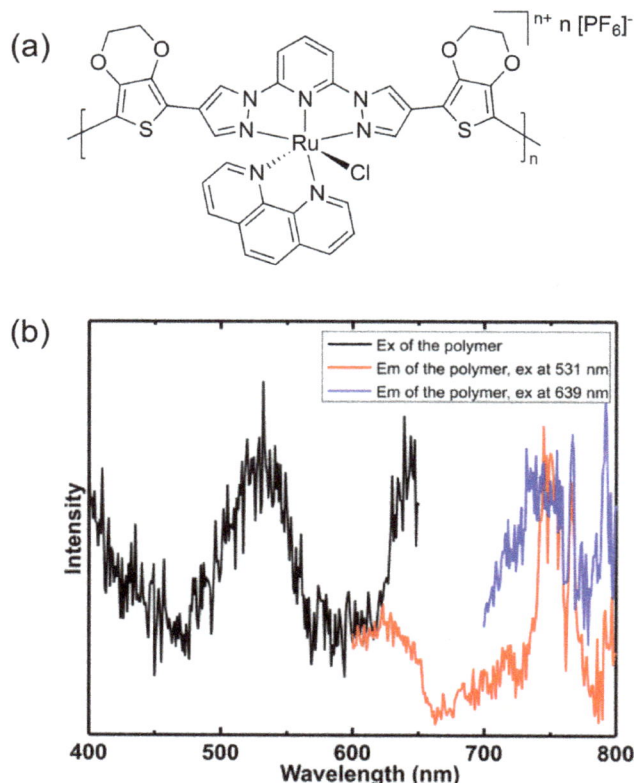

Figure 4.20 An example showing that the presence of a photoluminescent organometallic complex, tris(2,2'-bipyridyl)ruthenium(II), in (a) an organometallic polymer imparts (b) photoluminescence properties as revealed by the excitation and emission spectra of the polymer. Reproduced from ref. 31 with permission from American Chemical Society, Copyright 2018.

$$E_g^{opt} = \frac{hc}{\lambda_{a.e}} \tag{4.40}$$

Since we have discussed absorption spectroscopy, we should briefly mention transient absorption spectroscopy, which measures the change in absorbance as a function of time and wavelength to capture the transition of ground-state charge carriers to excited states. Although less commonly available in polymer laboratories, transient absorption spectroscopy is a powerful tool to characterise non-radiative relaxation processes and ultrafast transient charge carriers generated upon photoexcitation of photoactive polymers.

4.4.4 Electrochemical Methods

In Chapter 3, we introduced electrochemistry as a polymerisation method. The method is bifurcated, being used to characterise the redox activity of polymers as well. Many organometallic and coordination polymers undergo electrochemical processes involving metal- and/or ligand-centred redox processes studied by electrochemical methods. These methods can be classified into potentiometry, which measures the potential of a sample against a standard of known potential, and voltammetry, which applies a potential waveform to a sample and then measures the induced current response. Because of the limited capability of potentiometry in polymer characterisation, we will discuss voltammetry only, specifically cyclic voltammetry, which is readily available in most chemistry laboratories.

Cyclic voltammetry can probe the redox behaviour of a sample in 3D composed of current, potential and time. Theoretically, these data can be used to clarify if a sample is redox-active and under what conditions is electron transfer reversible or irreversible. The method also unravels kinetic and thermodynamic parameters, including the electron transfer mechanism, rate constants, diffusion coefficients, standard redox potentials, and electron stoichiometry, that describe the redox behaviour of the sample. From the electrochemical perspective, reversibility means the sample is oxidisable and reducible at potential close to the standard redox potential. Also, the electron transfer rate at the electrode surface is sufficiently fast at all potentials to ensure the process is not limited by electron transfer kinetics.

A typical cyclic voltammetry experimental setup includes three electrodes, namely, a working electrode, a counter electrode, and a reference electrode, immersed into the polymer solution, which usually includes a supporting electrolyte to increase solution conductivity. The potential applied to the working electrode changes linearly with time and is measured against the reference electrode maintained at a fixed potential, while the current response flows from the working electrode to the counter electrode. The results presented as a voltammogram is a plot of current response (I) as a function of applied potential (V) (Figure 4.21b). A complete cyclic voltammetry experiment consists of a forward sweep that records the maximum current response (I_p^{ox}) corresponding to potential, E_p^{ox}, associated with the oxidation process and a reverse sweep that measures the minimum current (I_p^{red}) corresponding to the potential, E_p^{red}, associated with the reduction process (Figure 4.21b).

An exemplary reversible one-electron redox process illustrated by that of ferrocene is shown in Figure 4.21;[33] the difference between E_p^{ox} and E_p^{red} equals 59 mV, but

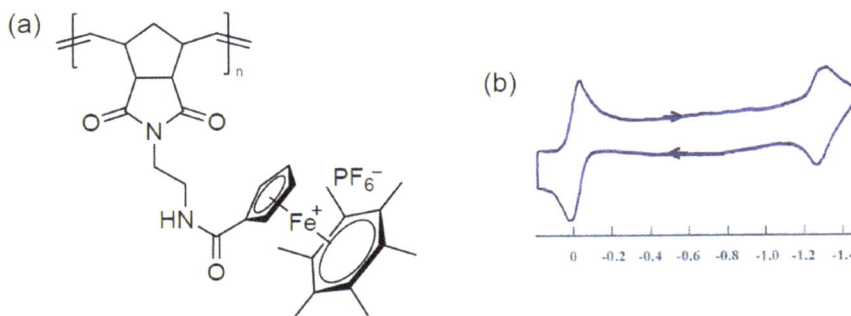

Figure 4.21 (a) A redox-active organometallic polymer characterised by cyclic voltammetry. (b) Cyclic voltammogram of the polymer in DMF. The internal reference is ferrocene, and the working and counter electrodes are platinum. The reversible redox wave at 0 V is that of ferrocene, while that at −1.3 V *vs.* ferrocene is that of the polymer. Reproduced from ref. 33 with permission from American Chemical Society, Copyright 2018.

experimental values approach 80 mV. In the reversible redox process ($I_p^{ox}/I_p^{red} = 1$), the number of electrons (z) involved in the electron transfer process is given by eqn (4.41).

$$\Delta E = E_p^{ox} - E_p^{red} \approx \frac{0.059 \text{ V}}{z} \tag{4.41}$$

The value of half-wave potential ($E_{1/2}$) for the redox process is reported against the reference electrode and obtained from eqn (4.42).

$$E_{1/2} = \frac{E_p^{ox} + E_p^{red}}{2} \tag{4.42}$$

In some organometallic and coordination polymers, the electrochemical process is quasi-reversible or irreversible ($I_p^{ox}/I_p^{red} \neq 1$) due to the slow electron transfer between the electrode and the redox couple or a chemical reaction accompanied by the electron transfer process. Irreversible and quasi-reversible redox processes are usually observed in some organometallic dendrimers containing different spheres of redox centres.[34] When a chemical reaction is accompanied by the electron transfer process, voltammetry provides a powerful method to investigate reaction kinetics.[35] Cyclic voltammetry may be used to compute the electrochemical bandgap, E_g^{ec}, of organometallic polymer-based semiconductors as given in eqn (4.43). The E_g^{ec} is calculated from the oxidation, $E_{p,onset}^{ox}$, and reduction, $E_{p,onset}^{red}$, onset potentials obtained from a cyclic voltammetry analysis of the polymer film cast on the working electrode.[36]

$$E_g^{ec} = \left(E_{p,onset}^{ox} - E_{p,onset}^{red}\right) \text{ eV} \tag{4.43}$$

4.4.5 Four-point Probe Conductivity Measurement

Measurement of electrical conductivity, the reciprocal of resistivity, is necessary for classifying materials as conductors, semiconductors, or insulators. Accurate determination of electrical conductivity in organometallic and coordination polymers is critical not just for classification purposes but also for comprehending and predicting their performance in electrical devices. The four-point probe conductivity method is the method of choice applicable to bulk samples or thin films and capable of providing accurate results. The method depends on the accurate measurement of the resistance and sample dimension. Powder polymer samples are usually homogenised and compressed into a pellet. For a pellet of length L and uniform cross-section area A, the resistance, which is the ratio of voltage to current, is related to resistivity by eqn (4.44). The typically small amount of current driven through a high impedance voltmeter ensures minimal cable and contact resistance effects on the results. The four-point probe method can measure resistance from 10^5 Ω down to 10^{-3} Ω in semiconductor surfaces when used with a micro-ohmmeter. Usually, the conductivity data are plotted as a function of temperature to classify the polymer as a semiconductor or metallic conductor, with the conductivity of semiconductors increasing and that of metals decreasing with temperature. The four-point probe method measures the room temperature electrical conductivity, which ranges from 10^{-4} to 5 Ω^{-1} cm^{-1}, in a series of organometallic polymers $(MC_2S_4{}^{x-})_n$ where M is Ni, Pd, Pt, or Au and shows that the temperature dependence of the conductivity of the Ni-containing polymer is similar to that of metals.[37]

$$R = \frac{\rho L}{A} \tag{4.44}$$

4.4.6 Magnetometry

Magnetometry is the measurement of the bulk magnetic moment of a sample. All materials have a magnetic moment, and in the case of organometallic polymers the most significant contributor to the magnetic moment is unpaired electrons. Magnetometry provides information on anisotropy, spin structures, and phase transitions for strongly magnetic materials. For paramagnetic materials exemplified by some organometallic polymers, measuring the total magnetic moment at a high field or the temperature dependence of magnetic moment susceptibility can reveal the degree of the magnetic moment per chemical constituent.

In most laboratories, the magnetic moment is directly measured by detecting the variation in magnetic flux produced by the sample or variation in the force experienced by the sample. These changes are transduced to electronic signals as direct current that is proportional to the magnetic moment. Transducers for flux-based detection are typically pickup coils, while those for force-based detection are electromechanical balances or piezoelectric devices. In flux-based magnetometers, which include extraction, vibrating sample, and superconducting quantum interference device (SQUID) magnetometers, a polymer sample is moved relative to the pickup coil that senses the change in flux. A change in flux can also be induced by changing the field strength or

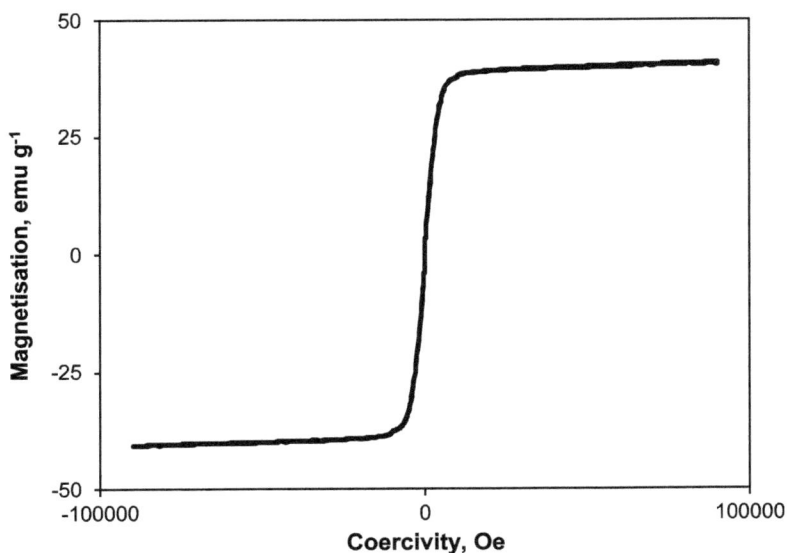

Figure 4.22 A hysteresis curve, *i.e.* a plot of magnetisation against field strength, coercivity, obtained with a vibrating sample magnetometer which measures magnetisation in a magnetic ceramic derived from the dendrimer in Figure 4.11. Reproduced from ref. 23 with permission from the Royal Society of Chemistry.

temperature. In contrast, a force-based magnetometer, such as a Faraday magnetometer, measures the energy difference induced in the sample due to a magnetic field gradient. Both flux- and force-based magnetometers measure the magnetisation as a function of temperature and magnetic field in organometallic polymers. Figure 4.22 shows a hysteresis curve, a plot of magnetisation against field strength, obtained with a vibrating sample magnetometer which measures magnetisation in a dendrimer-derived magnetic ceramic.[23]

4.4.7 Bioactivity Assays

Inorganic and organometallic polymers are increasingly investigated as biomaterials. Therefore, characterising their bioactivity and biocompatibility is critical to their use as biomaterials. The intended application informs the type of bioassay required to ascertain a biological property. For example, for a polymer intended for antimicrobial application, a disc-diffusion antibiotic susceptibility test can be performed, while a polymer-based anti-cancer agent will be characterised using an appropriate cytotoxicity assay. Cytotoxicity assays are bifurcated, being used to screen anti-cancer activity and to establish biocompatibility. To be specific, a compound generating a cytotoxic response against healthy cells or tissues is eliminated from further screening but considered a potential drug candidate for subsequent investigation if the response is directed towards rapidly proliferating cancer cells. The haemolytic assay, which evaluates the ability of a polymer to lyse red blood cells, is the gold standard biocompatibility assay if the biomaterials are intended for clinical application. The samples for biomedical application must be of high purity, usually greater than 95% purity, to confidently attribute activity to the desired compound.

Several bioassays exist to ascertain the mechanism of biological action. For instance, the bioactivity of some organometallic polymers, including redox-active and photoactive polymers, depends on the generation of reactive oxygen species (ROS).[14] Therefore, bioassays that monitor the levels of ROS or oxidative damage to biomolecules will elucidate the mechanism of action of these polymers. For example, the dichlorodihydrofluorescein-based oxidative stress assay confirms that the antimicrobial activity of the organometallic dendrimer in Figure 4.11 is due to the generation of ROS. The assay relies on the ROS driven oxidation of nonfluorescent dichlorodihydrofluorescein to highly fluorescent 2′,7′-dichlorofluorescein (Box 4.8).

Box 4.8 Review questions.

1. Describe a most appropriate characterisation method to determine the M_n, M_w and $Đ$ of a polycationic organometallic polymer.
2. Assuming you have successfully synthesised an organometallic polymer designed as an anti-fungal agent for clinical application, explain a critical bioassay that must be completed to prove its biocompatibility.
3. Explain the difference between DSC and DMA. What types of analysis and information do these methods provide?

4. Describe how you can characterise the molecular weight, the presence of cobalt, the presence of paramagnetic species, redox activity, and bioactivity of a cobalt-containing polycationic polymer.

5. You were unable to observe a signal in the ^{59}Co NMR spectrum of a cobalto-cenium-containing polymer. Explain the probable cause and how you can circumvent the problem to observe a signal.

6. Describe a method to characterise the degree of crystallinity and morphology of an organometallic polymer.

7. Assuming you carried out a DSC experiment on a semi-crystalline polymer, describe the expected transitions as you ramped up the temperature.

Further Reading

1. J. F. Rabek, *Experimental Methods in Polymer Chemistry: Physical Principles and Applications*, Johns Wiley & Son Ltd, Chichester, UK, 1980.
2. E. N. Kaufmann, *Characterization of Materials*, Wiley-Interscience, Hoboken, NJ, 2003.
3. J. E. Mark, H. R. Allcock and R. West, *Inorganic Polymers*, Oxford University Press Inc., New York, 2005.
4. R. D. Archer, *Inorganic and Organometallic Polymers*, Wiley-VCH, New York, 2001.
5. B. J. Hunt and M. I. James, *Polymer Characterisation*, Blackie Academic & Professional, London, 1993.

References

1. S. H. Rose and B. Block, *J. Polym. Sci., Polym. Chem.*, 1966, **4**, 583–592.
2. R. Puyenbroek, P. Werkman, B. Rousseeuw, E. Van der Drift and J. van de Grampel, *J. Inorg. Organomet. Polym.*, 1994, **4**, 289–299.
3. P. Wisian-Neilson and M. A. Schaefer, *Macromolecules*, 1989, **22**, 2003–2007.
4. R. D. Archer and B. Wang, *Inorg. Chem.*, 1990, **29**, 39–43.
5. H. Chen and R. D. Archer, *Macromolecules*, 1995, **28**, 1609–1617.
6. R. H. Lammertink, M. Hempenius, G. J. Vancso, M. Os, M. J. Beulen and D. Reinhoudt, *Chem. Commun.*, 1999, 359–360.
7. H. Gu, R. Ciganda, P. Castel, J. Ruiz and D. Astruc, *Macromolecules*, 2016, **49**, 4763–4773.
8. P. Cotts, R. Miller, P. Trefonas III, R. West and G. Fickes, *Macromolecules*, 1987, **20**, 1046–1052.
9. H. Braunschweig, C. J. Adams, T. Kupfer, I. Manners, R. M. Richardson and G. R. Whittell, *Angew. Chem., Int. Ed.*, 2008, **47**, 3826–3829.
10. C. U. Pittman Jr, J. Lai, D. Vanderpool, M. Good and R. Prado, *Macromolecules*, 1970, **3**, 746–754.
11. C. U. Pittman Jr, R. L. Voges and W. R. Jones, *Macromolecules*, 1971, **4**, 291–297.
12. S. Floquet, S. Brun, J.-F. Lemonnier, M. Henry, M.-A. Delsuc, Y. Prigent, E. Cadot and F. Taulelle, *J. Am. Chem. Soc.*, 2009, **131**, 17254–17259.
13. W. Li, H. Chung, C. Daeffler, J. A. Johnson and R. H. Grubbs, *Macromolecules*, 2012, **45**, 9595–9603.
14. A. S. Abd-El-Aziz, C. Agatemor, N. Etkin, D. P. Overy, M. Lanteigne, K. McQuillan and R. G. Kerr, *Biomacromolecules*, 2015, **16**, 3694–3703.
15. W.-M. Zhou and I. Tomita, *J. Inorg. Organomet. Polym. Mater.*, 2009, **19**, 113–117.
16. Y. Sha, M. A. Rahman, T. Zhu, Y. Cha, C. W. McAlister and C. Tang, *Chem. Sci.*, 2019, **10**, 9782–9787.
17. J. B. Gilroy, S. K. Patra, J. M. Mitchels, M. A. Winnik and I. Manners, *Angew. Chem., Int. Ed.*, 2011, **50**, 5851–5855.

18. R. Djeda, C. Ornelas, J. Ruiz and D. Astruc, *Inorg. Chem.*, 2010, **49**, 6085–6101.
19. A. S. Abd-El-Aziz, C. Agatemor, N. Etkin, R. Bissessur, D. Overy, M. Lanteigne, K. McQuillan and R. G. Kerr, *Macromol. Biosci.*, 2017, **17**, 1700020.
20. M. Shi, J. Yang, Y.-Y. Liu and J.-F. Ma, *Dyes Pigm.*, 2016, **129**, 109–120.
21. A. S. Abd-El-Aziz, C. Agatemor, N. Etkin and R. Bissessur, *Macromol. Chem. Phys.*, 2015, **216**, 369–379.
22. R. Prange and H. R. Allcock, *Macromolecules*, 1999, **32**, 6390–6392.
23. A. S. Abd-El-Aziz, C. Agatemor, N. Etkin and R. Bissessur, *J. Mater. Chem. C*, 2017, **5**, 2268–2281.
24. I. Gomez, D. Mantione, O. Leonet, J. A. Blazquez and D. Mecerreyes, *ChemElectroChem*, 2018, **5**, 260–265.
25. B. Ghanbari, L. Shahhoseini, A. Owczarzak, M. Kubicki, R. Kia and P. R. Raithby, *CrystEngComm*, 2018, **20**, 1783–1796.
26. M. Lippi, M. Cametti and J. Martí-Rujas, *Dalton Trans.*, 2019, **48**, 16756–16763.
27. B. Wunderlich, *Thermal Analysis of Polymeric Materials*, Springer-Verlag, Berlin Heidelberg, 2005.
28. R. Sumitani and T. Mochida, *Soft Matter*, 2020, **16**, 9946–9954.
29. A. A. Joraid, R. M. Okasha, M. A. Al-Maghrabi, T. H. Afifi, C. Agatemor and A. S. Abd-El-Aziz, *J. Inorg. Organomet. Polym. Mater.*, 2020, **30**, 1–15.
30. S. K. Petrovskii, A. V. Paderina, A. A. Sizova, A. Y. Baranov, A. A. Artem'ev, V. V. Sizov and E. V. Grachova, *Dalton Trans.*, 2020, **49**, 13430–13439.
31. S. G. Keskin, X. Zhu, X. Yang, A. H. Cowley and B. J. Holliday, *Macromolecules*, 2018, **51**, 8217–8228.
32. J. C. Costa, R. J. Taveira, C. F. Lima, A. Mendes and L. M. Santos, *Opt. Mater.*, 2016, **58**, 51–60.
33. H. Gu, R. Ciganda, R. Hernandez, P. Castel, P. Zhao, J. Ruiz and D. Astruc, *Macromolecules*, 2015, **48**, 6071–6076.
34. D. Astruc, *Nat. Chem.*, 2012, **4**, 255–267.
35. R. S. Nicholson and I. Shain, *Anal. Chem.*, 1964, **36**, 706–723.
36. Q. Sun, H. Wang, C. Yang and Y. Li, *J. Mater. Chem.*, 2003, **13**, 800–806.
37. R. Vicente, J. Ribas, P. Cassoux and L. Valade, *Synth. Met.*, 1986, **13**, 265–280.

5 Polymer Properties

5.1 Introduction

In Chapter 1, we saw how polymers contribute to technological development and the potential of inorganic and organometallic polymers to advance this development. In Chapters 2 and 3, we learned the various synthetic methodologies used in designing these polymers. Now, we will examine the properties that make them attractive materials. Generally, polymers are highly malleable, making them easily shaped. Also, they are relatively unreactive, allowing their utilisation under diverse conditions. The chemical inertness of polymers is an attractive property that enables their detailed characterisation, versatile processing, and broad application. However, unlike metals and ceramics, the thermal stability of polymers is unexceptional, as they tend to soften or decompose at moderate temperatures. Fortunately, the poor thermal stability and other properties can be tuned by controlling the molecular weight and structure. Other desirable properties such as electrical conductivity, magnetism, photoactivity, and redox activity, rare in the first-generation synthetic polymers such as polyethylene, are attainable through rational design of the polymer composition and structure. In this chapter, we will learn how the structure and composition of polymers are related to these properties, providing us with the knowledge to optimise the properties for specific applications (Box 5.1).

> **Box 5.1** Learning outcomes.
>
> By the end of this chapter, the student should:
>
> 1. Describe the different properties of polymers.
> 2. Describe structural trends that determine polymer properties.
> 3. Predict properties from structural and compositional features.
> 4. Identify functional properties.
> 5. Design polymers that exhibit desired properties.
> 6. Describe the thermodynamics of polymer solutions.
> 7. Describe the thermodynamics for experimental characterisation of thermal properties.

Fundamentals of Inorganic and Organometallic Polymer Science
By Christian Agatemor, Kajal Ghosal, Samuel Fura and Peter J. S. Foot
© Christian Agatemor, Kajal Ghosal, Samuel Fura and Peter J. S. Foot 2024
Published by the Royal Society of Chemistry, www.rsc.org

The term "property" in materials science describes a measurable stimulus-responsive material trait such as electrical conductivity and thermal stability. This definition excludes important macroscopic properties such as molecular weight and dispersity (discussed in Chapter 4). In this chapter, we will first examine the polymer–solvent interaction considering its influence on synthesis, characterisation, and processing. Then, we will discuss solid-state properties such as morphology and functional properties such as electrical conductivity and photoactivity. At the end of this chapter, we shall see that polymer properties can be tuned through control of the structure and composition. The knowledge gained will provide the student with a framework to predict and optimise a property and design a polymer with the desired property for a specific application.

5.2 Polymer–Solvent Interaction

Polymer–solvent interaction has a great influence on polymer syntheses (Chapters 2 and 3), characterisation (Chapter 4), processing and application (Box 5.2 and Chapter 6). When a polymer is in contact with a solvent, this interaction arises, disrupting the cohesive intermolecular force within the polymer matrix and ultimately resulting in normal dissolution or cracking. Unlike small molecules that dissolve instantaneously, polymer dissolution is a gradual process, involving chain disentanglement or diffusion through a

---

Box 5.2 Implications of polymer–solvent interactions for applications.

Polymer–solvent interactions are central to many polymer-based applications. We will discuss a few examples.

- *Polymer-based drug delivery systems.* These systems embed a drug within a polymer matrix. Some polymer matrices swell in a good solvent, allowing greater mobility and eventual diffusion/dissolution of the drug out of the matrix. A knowledge of polymer–solvent interactions helps in programming the drug release profile to achieve an optimum therapeutic response.
- *Polymer-based tissue engineering.* In tissue engineering, polymers can act as scaffolds, providing free volume and porosity to allow for cell proliferation and tissue vascularisation, and mechanical stability to support the implanted tissue. Polymer–solvent interactions initiate swelling, which affects the free volume and triggers a glassy-to-rubbery state transition, which influence the mechanical properties. A knowledge of these interactions is essential in selecting a suitable scaffold for tissue engineering.
- *Membrane science.* Polymer dissolution is exploited in membrane design. A microfiltration membrane, for example, can be formed from polymeric films cast from compatible pairs of polymers. Exposing the film to a selective solvent, dissolves one of the polymers, leaving interconnected microvoids to form a microfiltration membrane.
- *Plastic recycling.* In this application, knowledge of polymer–solvent interactions is vital in selecting a suitable solvent or solvent mixture to dissolve several plastics under the same or different conditions.

boundary adjacent to the polymer–solvent layer. The following subsections will provide the student with the basics of the dissolution process.

5.2.1 Mechanism of Polymer Dissolution

Polymer dissolution is an intensively investigated topic. However, most of the investigations used organic polymers, but some general considerations remain valid for inorganic and organometallic polymers. For example, the general understanding is that uncrosslinked amorphous polymer dissolution involves two critical transport processes: solvent diffusion into the polymer matrix, and chain disentanglement and migration into the pure solvent (Figure 5.1). Also, two mechanisms have been proposed to explain the dissolution of uncrosslinked amorphous polymers in contact with a thermodynamically compatible solvent. Both mechanisms suggest that solvent contact with the polymer results in a surface layer interface (Figure 5.1). In the first mechanism, termed normal dissolution, the interface consists of infiltration, solid swollen, gel, and liquid layers (Figure 5.1e). In the second mechanism, whereby the polymer cracks rather than dissolves, the gel layer is absent.

In normal dissolution, the polymer initially swells as solvent molecules infiltrate the free volume, then forms a gel-like layer adjacent to the swollen layer as the solvated polymer transitions from a glassy to a rubbery state (Figure 5.1). After an induction period, the polymer dissolves as the polymer chain disentangles and migrates into the pure solvent. If the polymer cracks rather than dissolves, the gel-like layer is understood to be lacking. For most solvent systems, the transition from normal dissolution to cracking depends on the gel temperature, the temperature at which the gel layer disappears, with polymers having low gel temperature dissolving in most solvents. Similar mechanisms explain the dissolution of semicrystalline polymers, except for an additional step where the crystalline domains first unfold into amorphous domains. In summary, polymers dissolve by forming a swollen gel-like layer, followed by chain disentanglement or extensive cracking.

5.2.2 Factors Controlling the Polymer Dissolution Rate

Several factors influence polymer dissolution. For a given solvent system and polymer, the molecular weight and dispersity play an immense role in the dissolution rate. Generally, the dissolution rate decreases with increasing molecular weight but increases with increasing dispersity. For example, in a series of bimetallic iron- and cobalt-containing organometallic dendrimers depicted in Figure 5.2, dissolution decreases with increasing dendrimer generation, which positively correlates with molecular weight.[1] The decreased dissolution rate precludes solution-state characterisation of these dendrimers at high generations.[1] The effect of molecular weight on dissolution is understandable considering that as the molecular weight increases, intramolecular forces and chain entanglement increase as well, requiring a higher degree of swelling to initiate dissolution. The inverse relationship between molecular weight and the dissolution rate is not indefinitely linear but levels off at very high molecular weights for some polymers.[2] Also, at very low molecular weights, the dissolution mechanism of some polymers transitions from normal dissolution to cracking.[2]

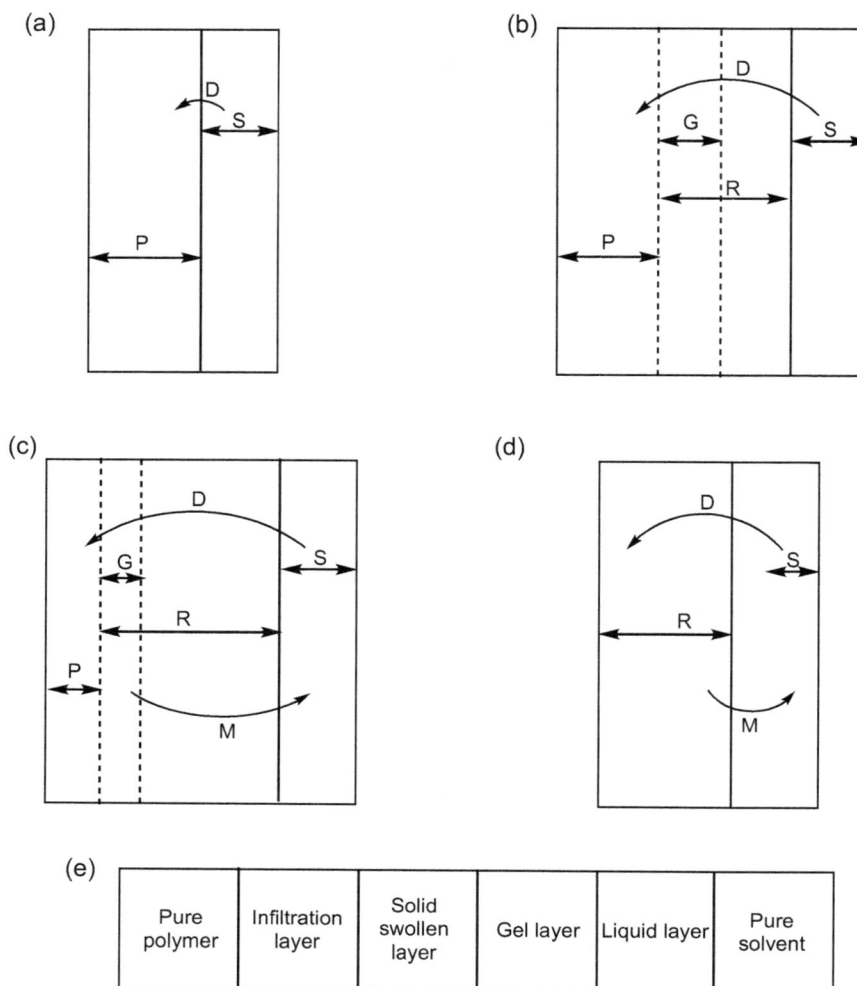

Figure 5.1 Schematic representation of the mechanism of polymer dissolution using a one-dimensional solvent diffusion and polymer dissolution model. (a) The initial slab of the glassy polymer (P) in contact with the solvent (S); (b) solvent diffusion (D) swells the polymer matrix and creates a glassy polymer–rubbery polymer interface (G) and a rubbery polymer–solvent interface (R); (c) the dissolution process starts as the positions of interfaces G and R decrease, and the polymer chain disentangles and migrates into the solvent; (d) the final dissolution step as interface G disappears, and transition to R, which gradually decreases as the polymer chain disentangles and migrates into the solvent; and (e) detailed composition of the interface.

Polymer composition and structure are the other factors that control the dissolution rate. Structural features such as the nature of the monomer, stereochemistry, and dimensionality (discussed in Chapter 1) profoundly affect the dissolution rate. Networked polymers are more likely to resist dissolution than their linear analogs, while copolymers dissolve more rapidly than homopolymers. Tacticity (discussed in Chapter 1) also plays a significant role in the dissolution rates of polymers. The impact of the structure on dissolution behaviour correlates with the glass transition temperature (T_g) (discussed in Section 5.2), with structures that increase T_g being more likely to decrease the dissolution

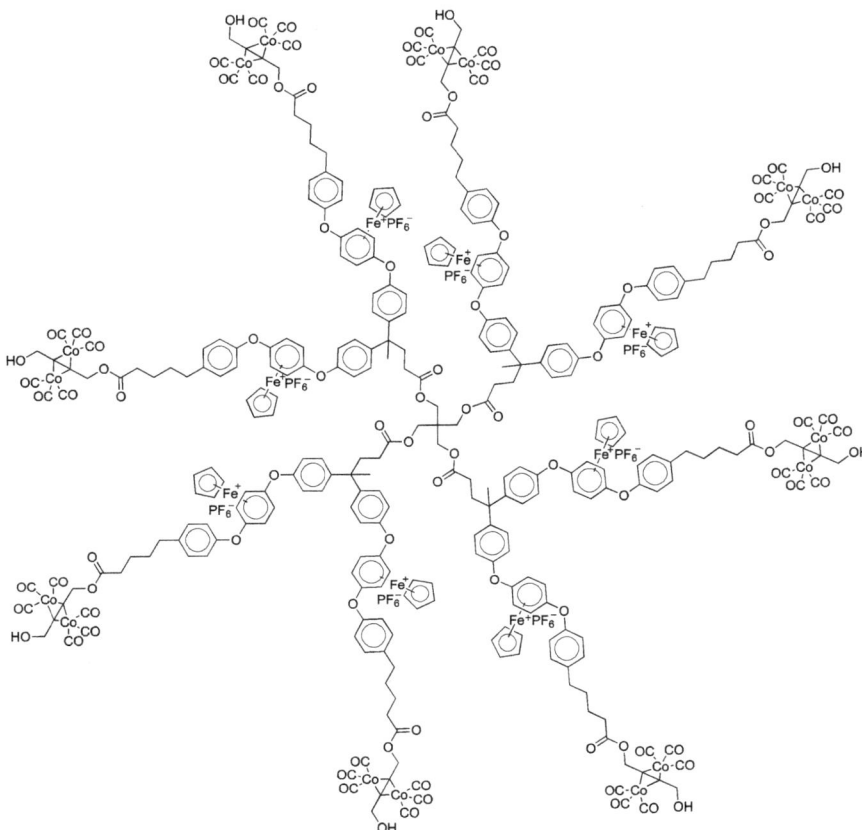

Figure 5.2 Schematic representation of the zeroth generation of a series of bimetallic iron- and cobalt-containing organometallic dendrimers. In this series, the dissolution rate of the polymer decreases with increasing dendrimer generation, which correlates with increasing molecular weight.

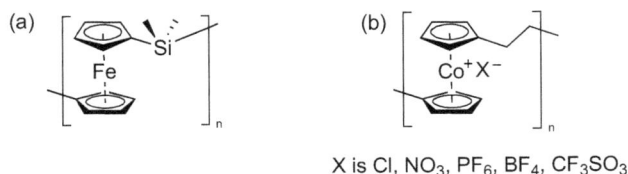

X is Cl, NO_3, PF_6, BF_4, CF_3SO_3

Figure 5.3 Schematic representations of a neutral and a cationic organometallic polymer. An ionic charge and the type of counteranion on the polymer significantly alter the dissolution profile.

rate. For example, the dendritic and hyperbranched structures feature lower T_g and dissolve more readily than their linear analogs that have relatively high T_g.

The impact of the polymer composition accounts for the differential dissolution behaviours of cationic *versus* neutral organometallic polymers. Charged species such as cobaltocenium cations enhance solubility in polar solvents, whereas neutral species such as ferrocene prefer non-polar solvents. For example, NMR spectroscopic characterisation of the neutral poly(ferrocenyldimethylsilane) (Figure 5.3a) is done in a non-polar solvent, such as benzene-d_6.[3] In contrast, that of the cationic poly(cobaltoceniumethylene) (Figure 5.3b) requires a polar solvent such as D_2O.[4] The different solubility behaviour of cationic and neutral organometallic polymers also informs the self-assembly (discussed in Section 5.3.3) of a copolymer containing blocks of neutral poly(ferrocenyldimethylsilane) and cationic poly(cobaltoceniumethylene).[5]

The counteranion in a polycationic polymer is another compositional parameter that affects dissolution behaviour. Cationic polymers containing weakly coordinating nucleophilic counteranions, such as tetrafluoroborate (BF_4) and hexafluorophosphates (PF_6), dissolve more readily in organic solvents, particularly polar ones, than those having strongly coordinating nucleophilic anions such as chloride and triflate. Even changing from weakly nucleophilic PF_6 to the relatively more nucleophilic BF_4 improves the dissolution of the cationic organometallic dendrimers in water.[6] On the other hand, switching from the hydrophilic triflate anion to an amphiphilic bis(2-ethylhexyl)sulfosuccinate anion enhances the dissolution of the copolymer poly(ferrocenyldimethylsilane)-*block*-poly(cobaltoceniumethylene) in organic solvents and contributes to its self-assembly properties.[5]

5.2.3 Polymer Solubility

Solubility, a thermodynamic process, differs from dissolution, although the two concepts involve solvents and solutes. Solubility is the maximum amount of solute that dissolves in a pure solvent under specific conditions. Solubility is, therefore, the endpoint of a kinetic process known as dissolution, which we just discussed in Section 5.2. The solubility of a polymer in a solvent depends on several factors but to a greater extent on the polymer structure and molecular forces holding the polymer together (discussed in Section 5.2.2). In line with the general principle of "*like dissolves like*", polymers dissolve in solvents with similar structures and inter molecular forces.

Detailed thermodynamic investigations show that the solubility of an amorphous polymer can be described by the relationship between the Gibbs free energy of mixing (ΔG_m), the enthalpy of mixing (ΔH_m), the entropy of mixing (ΔS_m), and absolute temperature (T) (eqn (5.1)).

$$\Delta G_m = \Delta H_m - T\Delta S_m \qquad (5.1)$$

ΔG_m is negative for polymers that dissolve spontaneously; otherwise, a multiphase system arises from the dissolution process. Also, polymer dissolution invariably results in a very small positive ΔS_m value due to an increase in the conformational mobility of the polymer chains in the solution. The small positive value of ΔS_m makes ΔH_m the fundamental parameter that determines the sign of ΔG_m. For a binary system, the enthalpy of mixing is related to concentration and energy parameters by eqn (5.2).[7]

$$\Delta H_m = V_{\text{mix}} \left[\left(\frac{\Delta E_1}{V_2} \right)^{\frac{1}{2}} - \left(\frac{\Delta E_2}{V_2} \right)^{\frac{1}{2}} \right]^2 \phi_1 \phi_2 \qquad (5.2)$$

V_{mix} is the total volume of the mixture; V_1 and V_2 are the molar volumes of components 1 and 2, respectively; E_1 and E_2 are the cohesive energies of components 1 and 2, respectively; $(\Delta E_1/V_2)$ and $(\Delta E_1/V_1)$ are the cohesive energy densities of components 1 and 2, respectively; and ϕ_1 and ϕ_2 are the volume fractions of components 1 and 2 in the mixture, respectively.

The cohesive energy density is the energy required to completely remove a molecule from its nearest neighbor, meaning to break all intermolecular forces in a unit volume of a material. At the same time, the square root of the cohesive energy density is the Hildebrand solubility parameter (δ). We can rewrite eqn (5.2) as eqn (5.3) to obtain the enthalpy of mixing per unit volume.

$$\frac{\Delta H_m}{V} = [\delta_1 - \delta_2]^2 \phi_1 \phi_2 \qquad (5.3)$$

For a polymer to dissolve over the volume fraction range ($\Delta G_m \leq 0$), ΔH_m must be smaller than ΔS_m (eqn (5.1)), which is achievable only if the difference in solubility parameter $(\delta_1 - \delta_2)$ is very small. In fact, if the difference is zero, solubility becomes entropy-driven. Therefore, achieving polymer dissolution requires the values of the Hildebrand solubility parameter for the solvent and polymer to be close in magnitude.

The requirement based on the Hildebrand solubility parameter is limited because it ignores contributions from the morphology, crosslinking, and specific interactions such as hydrogen bonds to polymer dissolution. Several chemists proposed mathematical models to overcome the limitations in the Hildebrand solubility parameter. Introducing hydrogen bonding into the solubility parameter, Burrell suggested that polymer–solvent miscibility is most favourable if the polarities of the polymer and solvent are similar, implying the need to match polymer and solvent polarities. Also, Hansen proposed to account for molecular interactions, including three specific interactions – dispersive interactions, dipole–dipole interactions, and hydrogen bonding interactions – in the cohesive energy (E) term of eqn (5.2). According to Hansen, cohesive energy comprises these three interactions as defined by eqn (5.4), used to obtain the cohesive energy density and solubility parameter as given in eqn (5.5) and (5.6), respectively.

$$E = E_D + E_P + E_H \qquad (5.4)$$

$$\frac{E}{V} = \frac{E_D}{V} + \frac{E_P}{V} + \frac{E_H}{V} \qquad (5.5)$$

$$\delta^2 = \delta_D^2 + \delta_P^2 + \delta_H^2 \qquad (5.6)$$

E_D, E_P, and E_H are the cohesive energy contributions from the dispersive interaction, dipole–dipole interaction, and hydrogen bonding interaction, respectively. Dividing the cohesive energy by the molar volume (V) gives the cohesive energy density and the Hansen solubility parameter. The values of Hansen solubility parameters for solvents and some inorganic polymers are available in the literature. Knowledge of the Hansen solubility parameter is very helpful to select the appropriate solvent or create a solvent mixture to dissolve a polymer.

5.2.3.1 Hydrodynamic Volume

A disentangled polymer chain in a solvent has several solution properties, such as hydrodynamic volume, with many practical implications. Hydrodynamic volume, defined as the effective volume or size of the polymer coil while in solution, can be used to determine the polymer size and molecular weight (discussed in Chapter 4). Hydrodynamic volume depends strongly on the polymer–solvent interactions, amongst other factors such as chain branching and the steric and electronic effects of the side-chain groups. The hydrodynamic volume increases with the polymer–solvent interaction; however, when this interaction decreases, intramolecular interactions become prevalent, contracting the volume.

$$r^2 = r_0{}^2 \alpha^2 \tag{5.7}$$

$$s^2 = s_0{}^2 \alpha^2 \tag{5.8}$$

A polymer chain forms a random coil conformation in solution, with the size of the chain describable by two dimensions, namely, the root-mean-square end-to-end distance, r^2, and the root-mean-square radius of gyration about the centre of gravity, s^2. While r^2 better defines the dimension of linear polymers, s^2 is more appropriate for branched polymers. These dimensions, r^2 and s^2, incorporate two parameters: the unperturbed dimension (r_0 or s_0) and an expansion factor (α), with the unperturbed dimension defining the size of the actual polymer random coil exclusive of the polymer–solvent interaction (eqn (5.7) and (5.8)). In contrast, the expansion factor results from the polymer–solvent interaction (eqn (5.7) and (5.8)). By rearranging eqn (5.7) and (5.8), the α parameters defined for mean-square end-to-end distance (r^2) and radius of gyration (s^2) become eqn (5.9) and (5.10), respectively. For a good solvent, defined as one that dissolves the polymer, α will be greater than unity because the perturbed r^2 and s^2 will typically exceed the unperturbed r_0 and s_0 dimensions. When α is unity, the polymer chain exists in the unperturbed configuration, behaving as ideal statistical coils.

$$\alpha = \frac{(r^2)^{\frac{1}{2}}}{(r_0{}^2)^{\frac{1}{2}}} \tag{5.9}$$

$$\alpha = \frac{(s^2)^{\frac{1}{2}}}{(s_0{}^2)^{\frac{1}{2}}} \tag{5.10}$$

The lowest temperature at which α is unity is known as the theta temperature, and the solvent is called a theta solvent. In a theta solvent, the polymer exists in the theta state, dissolved in the solvent but on the brink of insolubility. As the solvent changes from a theta to a poor solvent, the polymer–solvent interactions become unfavourable with the chain contracting and transitioning from a random coil to a globular shape conformation. The essence of the coil–globular transition is to minimise the polymer–solvent interactions and maximise the polymer intrachain and interchain interactions. Eventually, the concurrent increase in intrachain and interchain interactions leads to the formation of polymer aggregates in the solvent, which has implications for the light scattering-based molecular weight characterisation of polymers.

5.2.3.2 Viscosity

Even at low concentrations, polymer solutions are viscous, with the viscosity having an important implication for molecular weight determination. Indeed, the mean-square end-to-end distance, r^2 (discussed in Section 5.2.3.1), is related to the dilute solution viscosity and average molecular weight through eqn (5.11).

$$[\eta] = \frac{\phi(r^2)^{\frac{3}{2}}}{M} \tag{5.11}$$

$[\eta]$ is the intrinsic viscosity (discussed in Chapter 4), M is the average molecular weight, and ϕ is the Flory constant, roughly equal to 3×10^{24} mol^{-1}. The intrinsic viscosity is characteristic of a polymer and quantifies the increase in solution viscosity when the dissolved polymer concentration increases to a certain level. As expected, eqn (5.11) suggests that the intrinsic viscosity increases with the polymer dimension defined by the mean-square end-to-end distance. If we rewrite eqn (5.11) using eqn (5.7), we obtain eqn (5.12), which becomes eqn (5.13).

$$[\eta] = \frac{\phi(r_0^2\alpha^2)^{\frac{3}{2}}}{M} \tag{5.12}$$

$$[\eta] = \phi(r_0^2 M^{-1})^{\frac{3}{2}} M^{\frac{1}{2}}\alpha^3 \tag{5.13}$$

Given that ϕ, r_0 and M are constants for a given polymer with a defined molecular weight, it is possible to simplify eqn (5.13) into eqn (5.14).

$$[\eta] = KM^{\frac{1}{2}}\alpha^3 \tag{5.14}$$

K in eqn (5.14) is a constant equal to $\phi(r_0^2 M^{-1})^{\frac{3}{2}}$. At theta temperature, where α is unity, eqn (5.14) becomes eqn (5.15), while under other experimental conditions, it becomes eqn (5.16), known as the Mark–Houwink–Sakurada equation, which relates dilute solution viscosity with polymer molecular weight.

$$[\eta] = KM^{\frac{1}{2}} \tag{5.15}$$

$$[\eta] = KM^a \tag{5.16}$$

The parameters K and a in eqn (5.16) are polymer-, solvent- and temperature-dependent constants. For polymers composed of flexible chains dissolved in a good solvent, the value of a is around 0.7–0.8, while in a theta solvent it is 0.5. Polymers with rigid chains in a good solvent have values of a that exceed 1.0 due to their greater friction to mobility in the solvent.

5.2.3.3 Self-assembly

Upon dissolution in a selective solvent, polymers such as copolymers (Figure 5.4), comprising structurally incompatible segments, undergo self-assembly into well-ordered nanostructures above a critical aggregation concentration or critical aggregation temperature. This self-assembly behaviour results because the incompatible segments have different enthalpies and entropies of mixing with the solvent. This

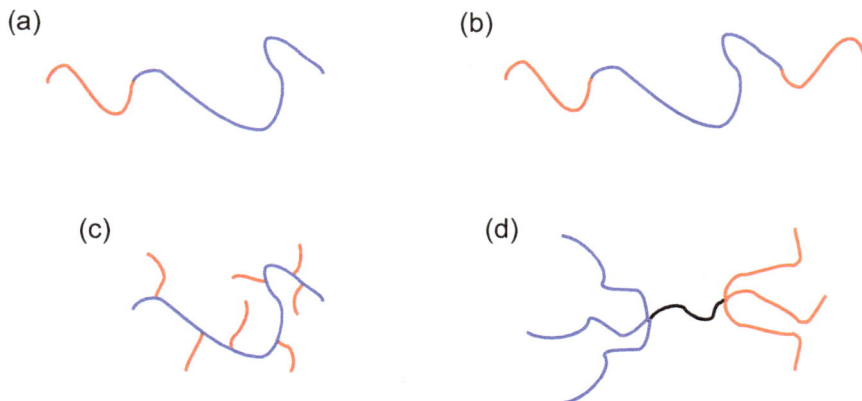

Figure 5.4 Some chain architectures of copolymers that are capable of self-assembly due to differential solubility of chain segments in a selective solvent: (a) AB linear block copolymer, (b) ABA linear block copolymer, (c) graft copolymer, and (d) Janus dendrimer.

difference drives the polymer chains to self-reorganise to minimise the unfavourable energetics due to a solvophobic segment–solvent interaction while preventing macroscopic phase separation of the segments. Solution self-assembly can occur at the single polymer chain level leading to single-chain folding or at the multichain level resulting in well-ordered nanostructures such as micelles and lamellae.

The Israelachvili packing parameter, p (eqn (5.17)), developed for the self-assembly of small molecule surfactants, can also predict the thermodynamically favoured nanostructure for a copolymer in a selective solvent.

$$p = \frac{v}{a_o l_c} \tag{5.17}$$

v is the solvophobic segment (tail) volume, a_o is the solvophilic segment interfacial area, and l_c is the solvophobic segment (tail) critical length (Figure 5.5). As a rule, spherical micellar nanostructures result when $p \leq \frac{1}{3}$, cylindrical micellar nanostructures such as nanorods when $\frac{1}{3} \leq p \geq \frac{1}{2}$, and bilayer membrane-like nanostructures such as vesicles or lamellae when $\frac{1}{2} \leq p \geq 1$ (Figure 5.5).

We must emphasise that Israelachvili's approach provides a simple yet powerful and intuitive model to predict the favoured nanostructures (Box 5.3). In some instances, the empirical results contradict the prediction due to the strong influence of the solvophobic segment on the interfacial area of the solvophilic segment. Nevertheless, we can engineer the nanostructures through control of the polymer chemical composition, dispersity of the constituent blocks, stereochemistry and crystallinity of the constituent blocks, and solvophilic/solvophobic block ratios defined by the molecular weight of the solvophilic block to that of the solvophobic block. For example, crystallinity dictates the nanostructure of poly(ferrocenylsilane-*block*-dimethylsiloxane) block copolymers. Poly(ferrocenyldimethylsilane-*block*-dimethylsiloxane) (Figure 5.6) with a crystalline poly(ferrocenyldimethylsilane) block forms a cylindrical micelle, while poly-(ferrocenylmethylphenylsilane-*block*-dimethylsiloxane) (Figure 5.6) with an amorphous poly(ferrocenylmethylphenylsilane) block forms a spherical micelle in hexane at

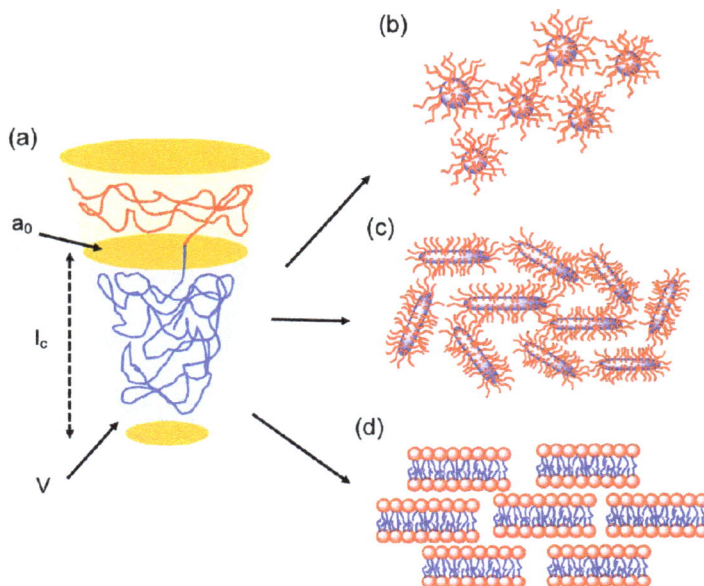

Figure 5.5 Examples of self-assembled nanostructures formed by amphiphilic copolymers in a block-selective solvent (a). The dimensionless packing parameter, *p*, obtained from eqn (5.17) determines the thermodynamically favoured nanostructure. Note: (b) $p \leq \frac{1}{3}$ favours the spherical micellar nanostructure form; (c) $\frac{1}{3} \leq p \geq \frac{1}{2}$ prefers cylindrical micellar nanostructures; and (d) $\frac{1}{2} \leq p \geq 1$ favours bilayer membrane-like nanostructures.

Box 5.3 Worked example 5.1.

Question

Use the Israelachvili packing parameter (*p*) to predict the thermodynamically favoured solution self-assembled nanostructure of the block copolymer depicted below.

a. When the degree of polymerisation (*m*) of the hydrophilic ethylene glycol block is very large relative to that (*n*) of the hydrophobic siloxane block.
b. When *m* is very small relative to *n* of the hydrophobic siloxane block.

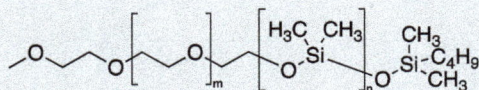

Answer

a. When *m* is very large, the interfacial area (a_0) will be large, resulting in a small *p* parameter (eqn (5.17)) and, therefore, spherical nanostructures such as micelles.
b. If *m* is very small, the interfacial area (a_0) will be small, resulting in a large *p* parameter (see eqn (5.17)) and, therefore, bilayer membrane-like nanostructures such as vesicles or lamellae.

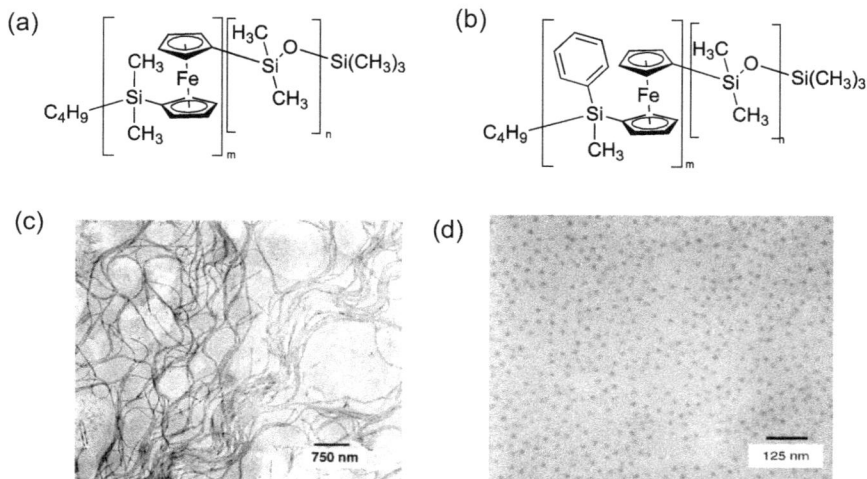

Figure 5.6 (a) Poly(ferrocenyldimethylsilane-*block*-dimethylsiloxane) with a crystalline poly(ferrocenyldimethylsilane) block forms (c) a cylindrical micelle in hexane. (b) Poly(ferrocenylmethylphenylsilane-*block*-dimethylsiloxane) with an amorphous poly(ferrocenylmethylphenylsilane) block forms (d) a spherical micelle in hexane. Reproduced from ref. 8 with permission from American Chemical Society, Copyright © 2000.

ambient temperature.[8] The macromolecular architecture (discussed in Chapter 1) and specific molecular interactions induced by diverse stimuli such as temperature, pH, and ions that influence the formation of irreversible covalent, dynamic covalent, and non-covalent bonds equally dictate the self-assembled nanostructure. It is important to note that self-assembly is not restricted to polymer solutions but occurs in the solid state as well. Self-assembly plays an important role in the solid-state dimensionality, morphology, and topology of polymers.

5.3 Solid-state Properties

We use polymers and their products primarily as solids. These solid polymers exhibit unique morphological, rheological, and thermal stability properties, which we will now consider.

5.3.1 Morphology

As solids, inorganic and organometallic polymers exhibit morphologies characteristic of crystalline, semicrystalline, or amorphous materials. The term amorphous or crystalline describes a physical state where the constituents of a material are randomly or regularly ordered, respectively. Highly crystalline polymers are rare, with only a few coordination polymers showing very high degrees of crystallinity that enable the growth of their single crystals for X-ray diffractometry (discussed in Chapter 5).[9] The majority of polymers are entirely amorphous or semicrystalline, with the latter possessing low to moderate degrees of crystallinity. Solid-state amorphous polymers like atactic poly(ferrocenylethylmethylsilane)[10] are strictly supercooled liquids, packed into space in their unperturbed dimensions as long, randomly coiled, interpenetrating polymer chains.

Polymer chains in the molten state engage in random micro-Brownian-like motion involving long polymer segments. As the melt cools, the glass transition temperature, T_g (discussed in Section 5.3.3), is reached where all long-range segmental motion stops. As the cooling continues below the T_g, only secondary relaxation processes that involve short-range motions such as rotations, vibrations, or flips of neighboring segments or side chains can occur. On cooling a crystalline polymer, exemplified by a family of lanthanide-containing polyoxometalates,[11] from the melt, the polymer chains fold into regular crystalline structures.

Several theories describe the morphology of the basic unit of crystalline polymers. The fringed-micelle theory proposes that polymers consist of ordered crystalline regions, called crystallites, imbedded in a disordered amorphous matrix (Figure 5.7a). The folded-chain lamella theory considers crystalline polymers as arrays of folded chains, with re-entry of each chain into the folded structure pictured as adjacent or non-adjacent. In the adjacent re-entry model, the chain forms tight, sharp, and regular folds as it re-enters the folded structure (Figure 5.7b). In the non-adjacent model, the chains coil through the surface of the lamella before re-entry into the folded structure (Figure 5.7c). The folded-chain lamella theory explains the morphology of single crystals of crystalline coordination polymers[11] and lamella of crystalline organometallic polymers such as poly(ferrocenyldimethylsilane).[12] Some polymers can crystallise into large spherical structures called spherulites which contain arrays of lamellar crystallites. We must re-emphasise that polymers are rarely crystalline; instead, most are amorphous, comprising only disordered domains, or semicrystalline, consisting of ordered and disordered domains (Box 5.4).

Whether a polymer will be semicrystalline or completely amorphous in the solid state depends on its chemical structure and intermolecular interactions within the polymer. Isotactic polymers with stereoregular structures favouring close packing have a higher degree of crystallinity than their atactic analogs. Polymers with symmetrical structures such as poly(ferrocenyldimethylsilane) pack more closely, resulting in a higher degree of crystallinity than those with unsymmetrical structures such as the completely amorphous poly(ferrocenylethylmethylsilane). Interchain interactions, such as hydrogen bonding, enable close packing into a crystal lattice and enhances the degree of

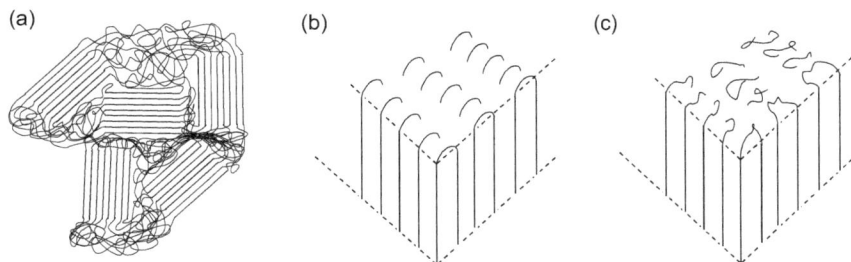

Figure 5.7 Different models of polymer crystallinity: (a) fringed-micelle theory, (b) adjacent folded-chain lamella theory and (c) non-adjacent folded-chain lamella theory.

> **Box 5.4** Understanding crystalline and amorphous polymers.
>
> **Amorphous Polymers**
>
> - They lack long-range molecular order.
> - They are unable to crystallise due to their composition. This may also describe crystallisable polymers (polymers whose composition permits partial crystallisation) that lack long-range molecular order under specific conditions.
> - They may contain localised order of about one nanometre length.
> - Provided stress is absent, the structure and properties are isotropic on a large scale.
> - Below the T_g, stressed amorphous polymers may be anisotropic, even after removing the applied stress.
>
> **Crystalline Polymers**
>
> - Wholly crystalline polymers are rare. Crystallisable polymers hardly crystallise completely. Crystalline domains coexist with amorphous domains, yielding semicrystalline polymers.
> - Three-dimensional, long-range order on an atomic scale characterises a significant fraction of semicrystalline polymers.
> - The degree of crystallinity describes the mass fraction or volume fraction of the crystalline domains.
> - The degree of crystallinity can be obtained by X-ray diffraction, differential scanning calorimetry, infrared spectroscopy, or solid-state nuclear magnetic resonance spectroscopy.

crystallinity even for atactic polymers. Extreme chain rigidity or flexibility precludes crystallisation. For example, the excessive flexibility of polysiloxanes prohibits close packing of the chains into a crystal lattice, making the polymer completely amorphous.

Morphology is a significant determinant of the mechanical properties, solubility, and thermal behaviour of a polymer. Generally, semicrystalline polymers are tougher and less soluble than their amorphous analogs. The enhanced mechanical properties and resistance to dissolution of semicrystalline polymers result from their tightly packed crystal lattices. The co-existence of crystalline and amorphous domains in semicrystalline polymers impacts their thermal behaviours. Typically, semicrystalline polymers exhibit a T_g arising from the amorphous domains and a crystalline melting temperature (T_m) due to the crystalline domains, whereas amorphous polymers exhibit only a T_g.

5.3.2 Rheology

Rheology deals with how materials deform and flow under stress. This topic is more of interest to physicists and engineers than chemists. Still, polymer chemists should grasp the fundamentals because of the pivotal role rheology plays in the behaviour of polymers during processing or application and to ensure seamless communication with physicists and engineers. For example, while it is acceptable in chemistry to describe amorphous polymers as liquids or solid materials depending on their T_g or T_m, it is more appropriate in physics to consider them viscoelastic materials at any temperature. Indeed, polymers behave as viscous liquids or elastic solids depending on the timescale of the deformation. To clarify, let us consider silly putty, a polysiloxane-based toy material, which fractures when subjected to a very rapid deformation by quickly pulling apart but bounces like an elastic material when subjected to a rapid deformation by dropping from a low height. In the latter case, rapidly deformed polysiloxane macromolecules revert to their original unperturbed, thermodynamically favourable conformation in a process called relaxation and store the energy, enabling bouncing. On the other hand, if the same silly putty is placed on the bench for a long time, the force of gravity makes it flow, behaving like a viscous material. This time-dependent viscoelasticity of materials like polysiloxanes has important implications for processing; the polymer can be shaped only by the slow application of force, whereas a very rapidly applied force fractures the polymer.

Whether a polymer is elastic or viscous depends on the timescale of the applied deformative force, more appropriately, the Deborah number (De), which is formally defined as the ratio of the relaxation time (λ) to the time of the deformative process (T) (eqn (5.18)). If De is low, the polymer exhibits predominantly Newtonian viscous behaviour; on the other hand, as De increases and becomes very large, the polymer enters the viscoelastic regime, and the rheology becomes increasingly dominated by Hookean elastic behaviour (Figure 5.8). The viscoelastic regime is of interest to polymer chemists and is divided into linear and nonlinear viscoelastic behaviours based on small and large deformative forces, respectively (Figure 5.8). Data obtained from viscoelasticity experiments provide information on molecular structure–rheology relationships to develop models for predicting polymer behaviour during processing and application. The goal of polymer chemists is to design a polymer structure for optimal performance during processing and application.

$$\text{De} = \frac{\lambda}{T} \tag{5.18}$$

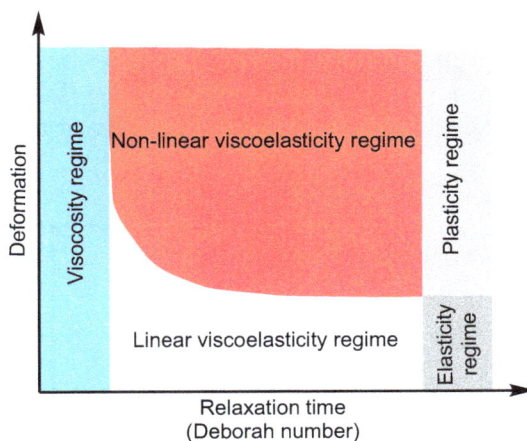

Figure 5.8 Schematic representation of viscosity, viscoelasticity, elasticity, and plasticity regimes as a function of deformation and relaxation times (Deborah number) during polymer deformation.

Various forces deform polymers, but academic research laboratories apply tangential stress, known as shear stress, to effect small deformations during rheological experiments. Shear stress (τ) is the applied force vector component parallel to the polymer cross-section. A cuboid-shaped polymer block subjected to shear stress deforms to a parallelepiped, provided the bottom of the block is held in place (Figure 5.9). In this case, we defined shear stress as the applied force per cross-sectional area parallel to the applied force (eqn (5.19)) and shear strain (γ) as the amount of deformation of the surface parallel to the applied shear stress (eqn (5.20)). Shear strain is measurable from the change in angle as the polymer deforms from a cuboid to a parallelepiped (Figure 5.9).

$$\tau = \frac{F}{A} \tag{5.19}$$

$$\gamma = \frac{X}{Y} = \tan\theta = \theta \; \text{ for small deformations} \tag{5.20}$$

The ratio of shear stress to shear strain, known as the shear modulus (G), describes a polymer response to shear. In the molten state, shear stress makes the polymer chains flow, and the susceptibility to flow is a function of temperature, molecular weight, molecular structure, and intermolecular interactions. Most polymer melts are non-Newtonian fluids, meaning their apparent viscosity is a function of the applied shear rate (the rate of application of the shearing deformation to the melt), the molecular weight and intermolecular interactions. The shear rate–shear stress relationship allows characterisation of time-independent fluids, whose flow behaviour depends only on the shear rate and is independent of shear history. A polymer melt exhibits Newtonian flow behaviour if shear stress increases proportionally with the shear rate, in agreement with Newton's law of viscosity (eqn (5.21)). Deviation from Newtonian behaviour predominates in most polymer melts. The most common time-independent non-Newtonian flow behaviour is shear thinning (Box 5.5), also known as pseudoplastic flow, when the apparent viscosity (η) gradually decreases with increasing shear rate. Conversely, shear thickening behaviour (Box 5.5), also known as dilatancy, results when apparent viscosity increases with the shear rate. Another deviation, though less common in polymer melts, is the Bingham Newtonian flow behaviour (Box 5.5) characterised by critical shear stress (τ_c), an initial resistance to flow that must be overcome before the melt flow as a Newtonian fluid pseudoplastic, or a dilatant (eqn (5.22)).

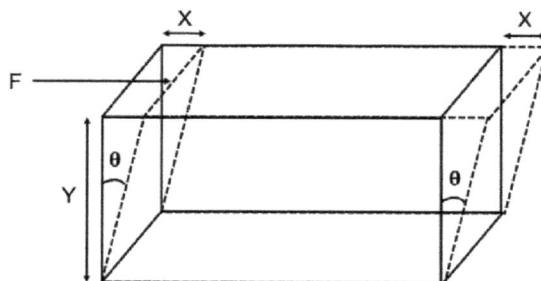

Figure 5.9 Illustration of an applied shear force (F) causing a displacement (X), deforming a rectangular block of a polymer to a parallelogram.

Box 5.5 Worked example 5.2.

Question

Provide a molecular-level explanation for the flow behaviour (Bingham, shear thinning and shear thickening) of polymers.

Answer

Before the polymer molecules flow, some conformation and molecular interactions probably need to be disrupted on applying the initial stress. The extent of this conformation and interactions are polymer structure- and composition-dependent, accounting for the different flow behaviours.

$$\tau = \eta\gamma \tag{5.21}$$

$$\tau = \tau_c + \eta\gamma \tag{5.22}$$

As we have mentioned, intermolecular entanglement, controlled by molecular weight, molecular structure, and intermolecular interactions, plays a pivotal role in rheology. For a polymer, a critical molecular weight (M_c) exists, below or above which the melt viscosity increases linearly or abruptly with molecular weight, respectively. For most polymers, the critical chain length that drives intermolecular entanglement to increase melt viscosity markedly corresponds to a degree of polymerisation (DP) of about 600. The DP also affects the yield stress, the amount of applied force that causes a change from elastic deformation to plastic deformation. For example, with some films of polydimethylsiloxane, the time for the yield stress to plateau increases with the DP, specifically from 9 minutes for a DP of 12 to 400 minutes for a DP of 85.[13] Also, some thick films of polydimethylsiloxane feature an enhanced viscous response compared to its liquid monomers.[13] Overall, the polymer chemist must consider the effect of molecular weight and intermolecular interactions on rheology during the design of a polymer. The molecular weight and degree of intermolecular interactions must be optimal, meaning "not too high" to ensure that melt viscosity does not preclude processing and fabrication and "not too low" to ensure adequate toughness and stability during use. We should remember that a dynamic mechanical analyser (discussed in Chapter 4) is used in most laboratories to characterise polymer rheology.

5.3.3 Thermal Transitions

A crystal of ice melting to liquid water on heating exemplifies a solid-to-liquid thermal phase transition. Polymers also undergo thermal phase transitions, with glass transition (T_g) and crystalline melting transition (T_m) being the most important from processing and application perspectives. The T_m, thermodynamically considered a first-order transition, is a unique property of the crystalline domains of a polymer. In contrast, the T_g, a second-order transition, is specific to the amorphous domains. Generally, thermal transitions are complex processes that are yet to be fully understood. Several theories classified as kinetic and equilibrium theories have been proposed to explain molecular level events that give rise to these transitions.

The kinetic theory considers these transitions as dynamic phenomena caused by "freezing" of the translational, rotational, and vibrational energies of polymer chain segments as the temperature decreases. To clarify, let us consider a polymer melt undergoing cooling. As the temperature drops, the translational, rotational, and vibrational energies of polymer chain segments decrease. Eventually, a temperature, T_m, is reached where translational and rotational energies become zero, permitting crystallisation if the polymer structure meets the appropriate symmetry requirements. Crystallisation is a kinetic process, mainly depending on nucleation, and usually occurs at a temperature lower than the T_m, a thermodynamic process. Some polymers, such as completely amorphous polymers, do not satisfy the symmetry requirement, precluding a tightly packed, ordered lattice arrangement of chain segments. Therefore, amorphous polymers lack a T_m. Whether a T_m exists or not, continuous cooling decreases the energies until a temperature, T_g, is reached where the segmental motion, which is the motion of the chain segment due to concerted bond rotation at the end of the segment, ceases. Below the T_g, bond rotation about side groups continues, even at extremely low temperatures in the glassy state.

Thermally induced changes in molecular level motions as described above alter various macroscopic properties of the polymer. For instance, a change in segmental motion at the T_g alters the free volume, leading to a concomitant change in specific volume. Therefore, measurement of the change in specific volume as obtained in dilatometry allows the determination of T_g. Other changes in macroscopic properties that can be measured to determine thermal transitions include the enthalpy change measured with DSC, and the stiffness or modulus change measured with a dynamic mechanical analyser (Box 5.6).

5.3.3.1 Thermodynamics of Thermal Transitions

A discussion of thermodynamics will help us grasp the basis for the experimental methods used to determine the thermal transitions of polymers.

Box 5.6   First- and second-order transitions.

A first-order transition exhibits a discontinuity in the first derivative of the Gibbs free energy (G) relative to some thermodynamic variables in contrast with a second-order transition defined by a discontinuity in the second derivative of G. With polymers, variables such as specific volume, entropy, and heat capacity change during thermal transitions. Specifically, during a crystalline melting transition, a discontinuous change in the density or enthalpy occurs at the transition temperature. Assuming a glass transition that is ideally a second-order transition, an abrupt, discontinuous change in heat capacity occurs at the transition temperature. However, the glass transition in polymers is a non-ideal thermodynamic transition because the discontinuity occurs gradually and is significantly affected by the timescale of the experiment. For example, fast cooling rates during DSC-based determination of glass transition result in high T_g values. Methods that measure these discontinuous thermodynamic variables, such as DSC which measures the change in heat capacity as a function of temperature, are used to determine T_m and T_g.

T_m. This first-order transition exhibits a discontinuity in the first derivative of the Gibbs free energy. For a reversible and closed system, the fundamental thermodynamic relationship between Gibbs free energy, G, molar volume, V, and molar entropy, S, can be expressed in the differential form as a function of temperature, T, and pressure, p, (eqn (5.23)).

$$dG = -SdT + Vdp \tag{5.23}$$

Differentiating G with respect to T at constant p yields eqn (5.24) and with respect to p at constant T yields eqn (5.25).

$$\left(\frac{\partial G}{\partial T}\right)_p = -S \tag{5.24}$$

$$\left(\frac{\partial G}{\partial p}\right)_T = V \tag{5.25}$$

Eqn (5.24) and (5.25) indicate that a first-order transition, as exemplified by the T_m of polymers, exhibits a discontinuous change in entropy at a constant temperature or a discontinuous change in volume at a constant pressure. For polymers, these equations provide a basis for the experimental determination of T_m, with eqn (5.24) being observable as a discontinuity in enthalpy and eqn (5.25) as a discontinuity in volume. Dilatometry can measure volume changes as a function of temperature at constant pressure, while calorimetry measures enthalpy changes. Figure 5.10 graphically depicts the determination of T_m from the thermodynamic first-order transition in volume as a function of temperature at constant pressure.

T_g. As we mentioned previously, the glass transition of polymers approximates a second-order transition, meaning it is discontinuous in the second derivatives of the Gibbs free energy. Eqn (5.26)–(5.28) depict three possible second derivatives of Gibbs free energy that provide a thermodynamic basis to determine the T_g experimentally.

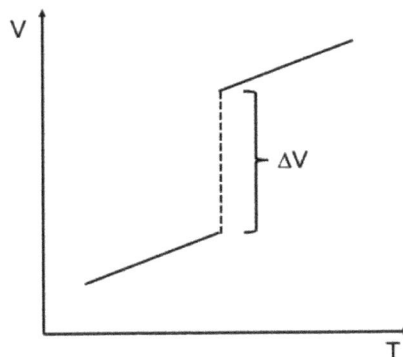

Figure 5.10 Thermodynamic first-order transition in volume as a function of temperature at constant pressure.

$$-\left(\frac{\partial^2 G}{\partial T^2}\right)_p = \left(\frac{\partial S}{\partial T}\right)_p \qquad (5.26)$$

$$\left(\frac{\partial^2 G}{\partial p^2}\right)_T = \left(\frac{\partial V}{\partial p}\right)_T \qquad (5.27)$$

$$\left[\frac{\partial}{\partial T}\left(\frac{\partial G}{\partial p}\right)_T\right]_p = \left(\frac{\partial V}{\partial T}\right)_p \qquad (5.28)$$

The equations indicate that the second-order transition should exhibit discontinuity in entropy as a function of temperature (eqn (5.26)), the slope of volume as a function of pressure (eqn (5.27)) or the slope of volume as a function of temperature (eqn (5.28)). Entropy is experimentally challenging to measure, so it is critical to relate eqn (5.26) with an easily measurable thermodynamic quantity. The quantity, specific heat at constant pressure, C_p, can be easily measured by calorimetry and is related to entropy through the first law of thermodynamics (eqn (5.29)). Substituting eqn (5.29) into eqn (5.26) yields eqn (5.30), implying that the second-order transition features a discontinuity in specific heat (Figure 5.11a).

$$C_p = T\left(\frac{\partial S}{\partial T}\right)_p \qquad (5.29)$$

$$-\left(\frac{\partial^2 G}{\partial T^2}\right)_p = \frac{C_p}{T} \qquad (5.30)$$

The discontinuities in slopes (eqn (5.27) and (5.28)) are related to the isothermal compressibility coefficient, β, and isobaric thermal expansion coefficient, α, through eqn (5.31) and (5.32). Substituting eqn (5.31) and (5.32) into eqn (5.27) and (5.28) yields eqn (5.33) and (5.34), indicating that a second-order transition occurs as a discontinuity in the isothermal compressibility coefficient and isobaric thermal expansion coefficient. Both coefficients can be measured by dilatometry. As we discussed previously, for an ideal second-order phase transition, C_p, α, and β change abruptly at the glass

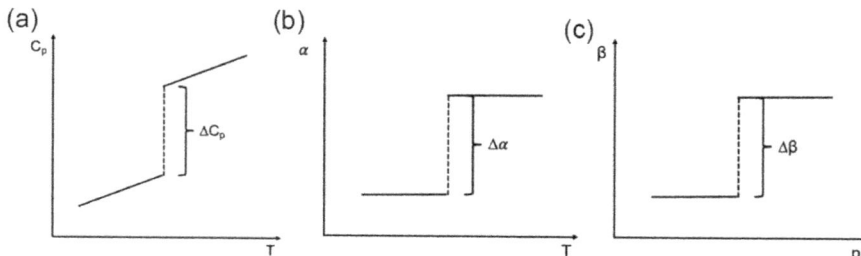

Figure 5.11 (a) Thermodynamic second-order transition in specific heat at constant pressure as a function of temperature. (b) Thermodynamic second-order transition in the isobaric thermal expansion coefficient as a function of temperature. (c) Thermodynamic second-order transition in the isothermal compressibility coefficient as a function of pressure.

Box 5.7 Worked example 5.3.

Question

Explain why glass transition in a polymer is considered a pseudo-second-order transition.

Answer

It means that if the glass transition is an ideal second-order transition, both the heat capacity C_p and the isothermal compressibility β (and the volumetric thermal expansion α) will change abruptly when the temperature reaches the glass transition temperature. However, for real polymeric systems, the change of these properties is gradual and affected by the heating rate.

transition temperature; however, for a real polymer, this change is gradual and depends on the heating rate (Figure 5.11) (Box 5.7).

$$\beta = -\left(\frac{1}{V}\right)\left(\frac{\partial V}{\partial p}\right)_T \tag{5.31}$$

$$\alpha = \left(\frac{1}{V}\right)\left(\frac{\partial V}{\partial T}\right)_p \tag{5.32}$$

$$\left(\frac{\partial^2 G}{\partial p^2}\right)_T = -V\beta \tag{5.33}$$

$$\left[\frac{\partial}{\partial T}\left(\frac{\partial G}{\partial p}\right)_T\right]_p = V\alpha \tag{5.34}$$

5.3.3.2 *Factors That Affect Thermal Transitions*

From the thermodynamic perspective, a negative Gibbs free energy is necessary for a spontaneous phase change. For a crystalline polymer below the equilibrium melting point, T_m^∞, crystallisation occurs spontaneously, whereas above the T_m^∞, melting occurs spontaneously. At T_m^∞, the crystalline and melt states are in equilibrium with the change in Gibbs free energy, ΔG, being zero. At the equilibrium melting point, the first law of thermodynamics indicates that the ΔG of fusion per polymer repeat unit is related to T_m^∞ as shown by eqn (5.35). Considering that ΔG is zero, we can define T_m^∞ by eqn (5.36).

$$\Delta G_f = \Delta H_f - T_m^\infty \Delta S_f \tag{5.35}$$

$$T_m^\infty = \frac{\Delta H_f}{\Delta S_f} \tag{5.36}$$

Eqn (5.36) implies that T_m^∞ depends on enthalpic and entropic factors. For example, the change in the entropy of the polymer melt due to changes in the polymer structure, such as increasing intermolecular interactions or chain stiffness, increases T_m^∞.

Although the observed T_m is always lower than T_m^∞, enthalpic and entropic considerations remain pivotal to this thermal transition.

Generally, the polymer structure similarly influences the two thermal transitions. It is rational to expect polymers with rigid structures that limit concerted bond rotation at the end of a chain segment, a hallmarked phenomenon during glass transition, to require high temperatures to induce the translational and rotational motions that occur during melting. On the other hand, flexible structures that readily permit concerted bond rotation at the end of a chain segment should require lower temperatures to undergo translational and rotational motions. Indeed, polymers with low T_g typically have low T_m, while those with high T_m feature high T_g. To illustrate, a flexible polymer, polydimethylsiloxane,[14] with a T_g of about −123 °C has a T_m of approximately −37 °C, while a more rigid polymer, polyferrocenylsilane,[15] with a T_g of about 30 °C has a T_m of about 132 °C. Overall, T_g and T_m depend greatly on chain flexibility as exemplified by the dramatic increase in T_m by 153 °C after replacing a flexible aliphatic group such as the methyl group in polydimethylsiloxane with a rigid aromatic group such as a phenyl ring in polydiphenylsiloxane.[14]

Chain flexibility, molecular symmetry, and intermolecular interactions are critical to the two thermal transitions. Flexible molecular structures such as silicon–oxygen–silicon bonds undergo molecular motions even at very low temperatures, reducing the values of T_g and T_m. Polysiloxanes feature very low values of T_m and T_g, so incorporating a polysiloxane into a polymer network is a strategy to lower the T_g of copolymers provided no phase separation occurs between the constituent blocks of the copolymers. Conversely, bulky substituents such as ferrocenes limit molecular motions, increasing T_g and T_m. Ferrocene-containing polymers exhibit very high T_g and T_m due to the bulkiness of ferrocene, which hinders chain rotation and symmetry, which promotes crystallisation. An example is the T_m of a polyamide that increases from 60 °C to 120 °C upon conjugation of ferrocene into the polymer backbone.[16] Also, cyclic structures such as aromatic rings reinforce polymer rigidity, increasing the values of T_g and T_m. The effect of the cyclic structure undoubtedly contributes to the dramatic difference in the T_g of polydimethylsiloxane, which is about −123 °C,[14] and that of polydiphenylsiloxane, which is about 40 °C.[17]

Another structural consideration is the length, number, and identity of side groups on the polymer backbone. Generally, these structural effects are difficult to predict, but the consensual hypothesis is that whatever restricts molecular motion increases T_g and T_m. The T_g and T_m can increase or decrease as the side group increases in length depending on whether the increasing length induces chain entanglement or side-chain crystallisation, or exerts a plasticiser effect. Structural effects such as branching and cross-linking significantly increase T_g. Molecular weight is another fundamental property that exerts a dramatic influence on T_g. Increasing the molecular weight reduces the number of chain ends and ultimately lowers the free volume. For example, the T_g of an organometallic dendrimer increases with dendrimer generation, which correlates with molecular weight.[18]

Increasing the attractive intermolecular interactions, such as hydrogen bonding, reduces the molecular mobility of amorphous polymer chains, increasing T_g. Also, increased attractive interactions promote crystallisation, increasing T_m. Attractive intermolecular interactions explain why the T_g and T_m of a ferrocene-containing polyamide ($T_g = 65$ °C and $T_m = 120$ °C) are higher than those of its polyester analog ($T_g = 50$ °C and $T_m = 75$ °C).[16] The hydrogen bonding between amide linkages contributes to the enhanced thermal properties of the polyamide (Boxes 5.8 and 5.9).

Box 5.8 Worked example 5.4.

Question

Use the T_g and T_m values in the table below to predict polymers W, X, Y, and Z from the figure below. Provide a rationale for your prediction.

Polymer	T_g (°C)	T_m (°C)
W	−66	260
X	−2	370
Y	4	390
Z	24	420

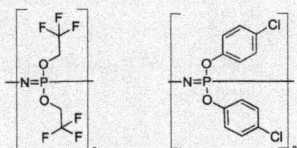

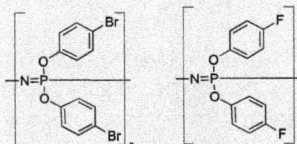

Answer

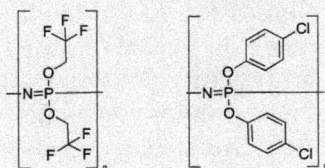

Polymer W
$T_g = -66$ °C
$T_m = 260$ °C

Polymer Y
$T_g = 4$ °C
$T_m = 390$ °C

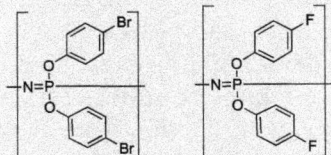

Polymer Z
$T_g = 24$ °C
$T_m = 420$ °C

Polymer X
$T_g = -2$ °C
$T_m = 370$ °C

Rationale: an aromatic ring impacts rigidity, restricting chain mobility and, ultimately, increasing T_g and T_m. Halogens can introduce halogen bonding in the polymer network. Halogen bonding is an attractive intermolecular force that increases in the order of F < Cl < Br. The student should refer to Young *et al.*, *Polymer*, 1992, **33**, 3215–3225.

Box 5.9   Worked example 5.5.

Is there a T_g or T_m that is appropriate for all applications and processing conditions?

No. An appropriate T_g or T_m depends on the application or processing temperature where the polymer will be used or processed. For example, a polymer intended for making coffee cups must have a T_g well above the temperature of hot coffee. Therefore, in tuning the thermal properties through control of the molecular structure, symmetry or intermolecular interactions, the chemist must consider the intended use and processing conditions of the polymer.

5.4   Functional Properties

Functional properties result from functional groups or structural motifs in the polymer framework. Therefore, only polymers carrying certain stimulus-responsive functional groups or structural motifs exhibit these properties, which are also pertinent for designing functional materials. This section will discuss the functional properties found in inorganic and organometallic polymers.

5.4.1   Liquid Crystalline Polymers

Liquids are isotropic and lack molecular order, whereas crystalline solids are anisotropic and exhibit molecular order. A liquid crystal displays properties of both liquids and crystalline solids, flowing like a liquid but having some anisotropic ordered systems. We mentioned in the previous sections that a polymer melt or concentrated solution consists essentially of completely disordered and randomly entangled chains. However, polymers can form ordered liquids. Indeed, in 1956, Flory suggested that, depending on the temperature range and the axial ratio, length-to-diameter ratio, of the polymer chain, quasi-crystalline arrangement of the chains in the liquid state can occur above a critical concentration. This arrangement is driven primarily by chain rigidity with attractive intermolecular interactions playing only minor roles.[19] It is now recognised that an essential criterion for liquid crystallinity is the presence of long, rigid, and highly anisotropic molecular structures, known as mesogenic groups or mesogens.

A liquid crystalline polymer incorporates the mesogenic group along its backbone or as a pendant to the backbone. Aromatic polyesters and polyamides typically exhibit liquid crystalline behaviour, as do many inorganic and organometallic polyesters and polyamides.[16] Also, some organometallic complexes such as ferrocenyl groups enhance liquid crystalline behaviour; specifically, they increase the clearing point of liquid crystalline polyamides (see Box 5.10 for the definition of the clearing point). Metal-containing mesogens incorporated within a polymer introduces rigidity and shape anisotropism, imparting liquid crystalline behaviour into polymers. For example, a polymer containing a copper(II) atom (Figure 5.12a) in the square planar geometry functions as a mesogen, resulting in several thermotropic, nematic coordination polymers.[20] Also, an organometallic complex, 1,3-diethynylcyclobutadiene-(cyclopentadienyl)cobalt (Figure 5.13), is a mesogen that enables the synthesis of an

Box 5.10 Types of liquid crystalline polymers.

- *Thermotropic*. These polymers melt to the liquid crystalline state at the melting point, then clear to the isotropic liquid state at high temperature. The transition from the crystal to the liquid crystalline phase, also known as the mesophase, is termed the melting point, while that from the mesophase to isotropic liquid is termed the clearing point.
- *Lyotropic*. These polymers require a solvent to form a liquid crystalline state; intuitively, the liquid crystallinity of lyotropic polymers depends on concentration and temperature.

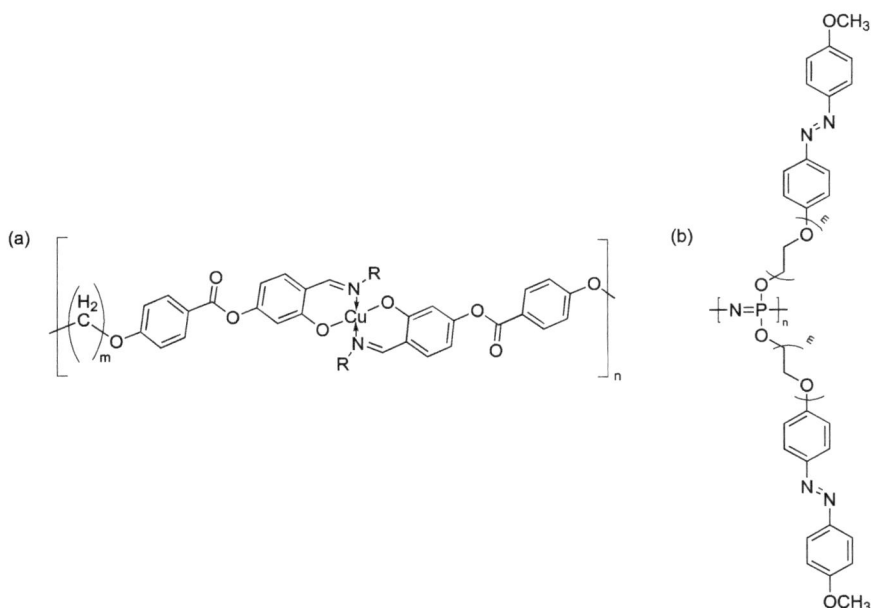

Figure 5.12 (a) A copper-containing liquid crystalline polymer. (b) A liquid crystalline polyphosphazene.

organometallic polymer with lyotropic and thermotropic nematic liquid crystalline behaviours.[21] Flexible inorganic polymers such as a polyphosphazene (Figure 5.12b) bearing rigid side-chain mesogenic groups can exhibit liquid crystallinity, provided the flexible backbone does not prohibit the ordering of the rigid mesogenic groups, and a flexible spacer links the rigid mesogenic group to the flexible backbone. The spacer is critical to the development of liquid crystallinity because it decouples the thermal motion of the backbone from the mesogenic group. When the spacer is absent or too short, liquid crystallinity disappears, as observed in the polyphosphazene in Figure 5.12b.[22]

Liquid crystalline behaviour affects other polymer properties such as viscosity and ordered orientation in the solid state. For example, the copper-containing liquid crystalline polymer in Figure 5.12a shows a dramatic decrease in viscosity at the

(b)

(a)

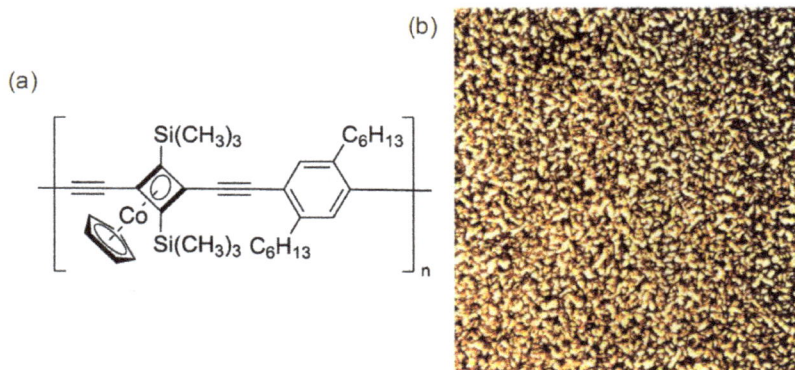

Figure 5.13 (a) A liquid crystalline polymer derived from an iron-containing sandwich complex. (b) A Schlieren texture of the liquid crystalline polymer at 165 °C. Reproduced from ref. 21 with permission from John Wiley and Sons, Copyright 1995 by VCH Verlagsgesellschaft mbH, Germany.

clearing point where it transitions from the isotropic liquid to the liquid crystalline phase.[20] Also, the schlieren texture of the organometallic polymer in Figure 5.13 remains after solvent evaporation, indicating a liquid crystalline behaviour in the solid state.[21]

5.4.2   Ionic Polymers

Ionic polymers incorporating ionic species within their framework as backbone structures or as pendant groups such as the dendrimer in Figure 5.2 are a unique class of functional polymers. These polymers can be obtained through chemical modification of neutral polymers or polymerisation of ionic monomers. In the chemical modification approach, a metal ion acts as a counterion to an ionic polymer, or a neutral polymer undergoes a reaction to form an ionic polymer (Figure 5.14). Typical examples are the syntheses of poly[(disodium carboxylatoethylphenoxy)phosphazene] by alkaline hydrolysis of poly[(dimethyl carboxylatoethylphenoxy)phosphazene] with sodium hydroxide[23] or oxidation of a neutral polyvinylferrocene to poly(vinyl-ferrocenium chloride) by iron(III) chloride (Figure 5.14).[24] In the other approach, an ionic monomer polymerises into an ionic polymer or a metal complex coordinates with an organic ligand to form an ionic coordination polymer. This approach is exemplified by the polymerisation of the cationic η^6-dichlorobenzene–η^5-cyclopentadienyliron(II) hexafluorophosphate to form a cationic iron-containing poly(aromatic ether)[25] or a transition metal complex such as iron(II) chloride with a terpyridine-functionalised ligand to form a coordination polymer (Figure 5.15).[26] Irrespective of the synthesis route, the resulting ionic polymer is classified as an ionomer or a polyelectrolyte depending on the ionic content. Ionic polymers with less than 15% ionic content are called ionomers, while those with higher ionic content are called polyelectrolytes.

Generally, ionic polymers consist of ionic groups and non-ionic polymeric segments. Incompatibility between the ionic and non-ionic segments results in micro-phase separation or self-assembly (discussed in Section 5.2), a characteristic of ionic polymers. In the presence of an oppositely charged substrate, the ionic groups can undergo

(a)

(b)

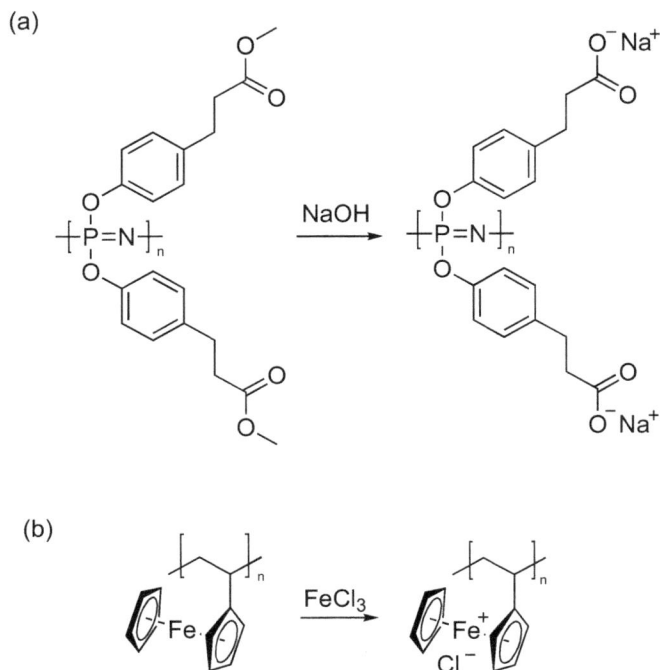

Figure 5.14 Chemical modification of neutral polymers to ionic polymers.

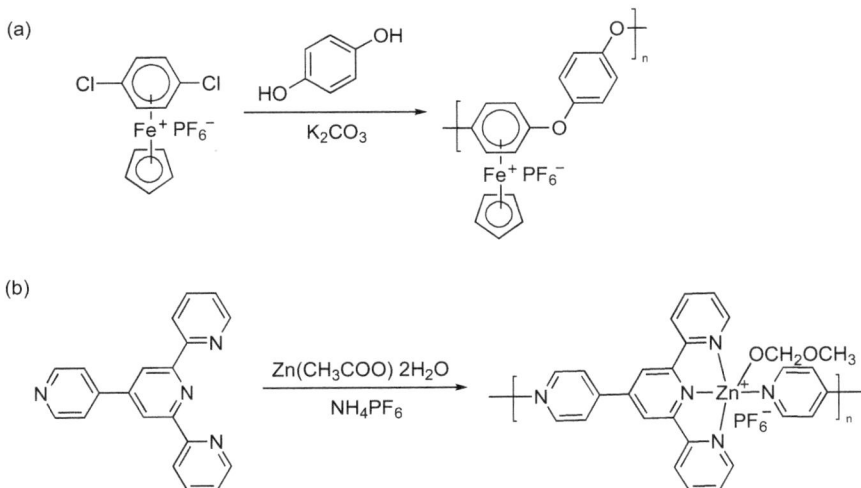

(a)

(b)

Figure 5.15 (a) The polymerisation of an ionic monomer to an ionic polymer. (b) Metal coordination of organic ligands to form ionic coordination polymers.

thermodynamically driven electrostatic interactions, as is observed with some cationic polymers that electrostatically interact with negatively charged cell membranes. The ionic groups significantly alter the polymer properties, such as the solubility, melt viscosity, modulus, and T_g. The degree of alteration is influenced by the type of ionic group and counterion, the position of the ionic group within the polymer framework, and the ionic content. The effect on solubility is evidenced by the enhanced aqueous solubility of ionic ferrocenium-based polymers derived by oxidation of neutral ferrocene-containing polymers (Figure 5.14b). Many ionic organometallic and coordination polymers are redox-active, a property harnessed to design stimulus-responsive, self-assembled, electrochromic, and self-healing materials. Their redox activity and ionic character also have implications for bioactivity, enabling anticancer and anti-pathogenic activities. Also, these polymers are promising materials for gene delivery and membrane fabrication.

5.4.3 Conjugated Polymers

Conjugated polymers are widely investigated as precursors to functional materials due to their peculiar photoactivity, magnetism, and electronic conductivity. These polymers are explored as sensors, photovoltaics, electrochromic displays, nonlinear optical materials, variable-transmission windows, and polymeric electrodes. Several families of organometallic and coordination polymers containing conjugated π-bonds, exemplified by polymetallalynes, poly(metallophthalocyanine)s, poly(metal tetrathiooxalate)s, polydecker sandwich complexes, metal poly(benzodithiolene)s (Figure 5.16), and cyclobutadienecobalt complex polymers (Figure 5.13), are electrically conductive or

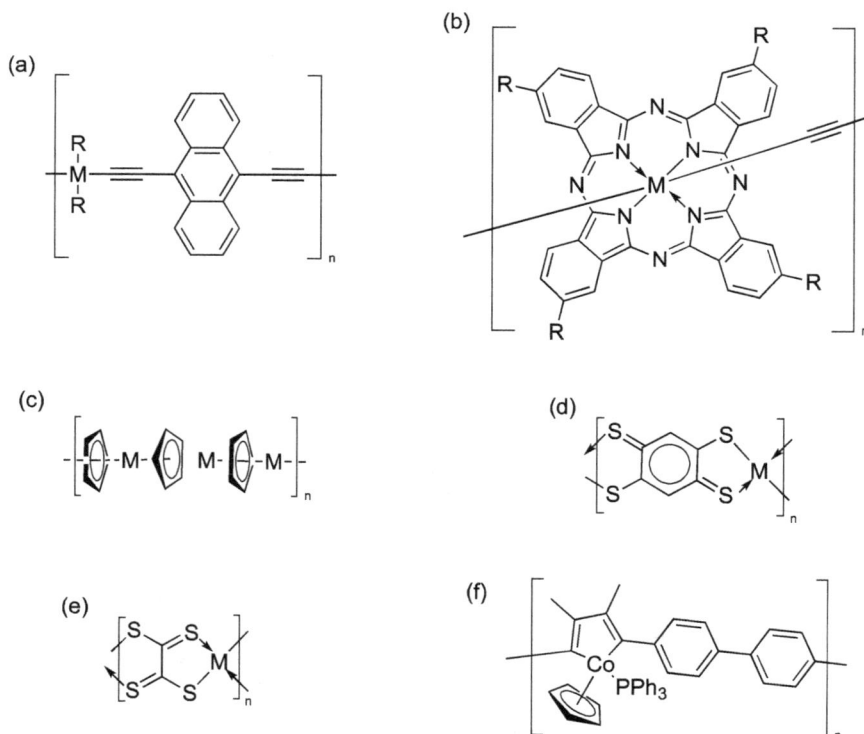

Figure 5.16 Examples of electrically conductive organometallic and coordination polymers. M represents the transition metal in the polymer structure. (a) Polymetallalynes, (b) poly(metallophthalocyanines), (c) polydecker sandwich complexes, (d) metal poly(benzodithiolenes), (e) poly(metal tetrathiooxalates), and (f) cobaltacyclopentadiene-based conductive polymers.

photoactive. Generally, the conjugated system has an unpaired electron, the π electron, per carbon atom that engages in π-bonding. A typical π-bond has an sp^2p_z electronic configuration. The unpaired electron resides on the p_z orbitals, which overlap in a polymer to form a molecular level electron delocalisation along the backbone. This electronic delocalisation is the pathway to the characteristic charge mobility in conductive polymers. The insertion of a transition metal into the conjugated system may increase electronic conductivity due to the unique electronic structure of the metal. Transition metals are electron-rich, thus contributing to the electron transfer dynamics within the polymer, and have d-orbitals that increase the likelihood of molecular orbital interactions. Together, both properties enhance electronic delocalisation within the polymer.

Chain symmetry, the number and identity of atoms in the polymer repeat unit, determines the electronic structure of conductive polymers, which can be semi-conducting or metallic in conductivity depending on the energy gap, E_g. The energy gap is the energy difference between the highest occupied state in the π band and the lowest unoccupied state in the π^* band.[27] Doping with an electron donor or acceptor improves the electrical conductivity of these polymers. As an example, the conductivity of the cobaltacyclopentadiene-based conductive polymer (Figure 5.16f) increases by 13 orders of magnitude, from 10^{-6} to 10^{-4} S cm^{-1}, upon doping with iodine.[28] The metal centre can contribute to the enhanced conductivity upon doping by generating a Co^{3+}/Co^{4+} mixed-valence state which interacts with the π-conjugated system, causing a mixed-valence conductivity. We must mention that the four-point probe conductivity test discussed in Chapter 4 is used to measure the electrical conductivity of these polymers in the solid state.

A transition metal inserted in the π-conjugated system of a polymer impacts and influences the overall redox (discussed in Section 5.4.4) electronic and photophysical properties. Usually, the transition metal improves the spin–orbit coupling, enhancing intersystem crossing from the singlet to the triplet state and increasing the triplet excited state population in conjugated organometallic polymers. Therefore, these polymers are also photoluminescent. We use a Jablonski diagram (Figure 5.17) to

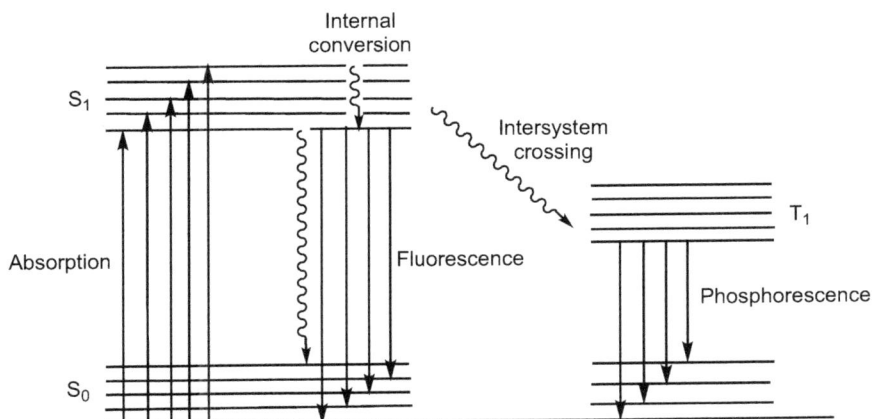

Figure 5.17 Jablonski diagram showing processes that lead to fluorescence and phosphorescence.

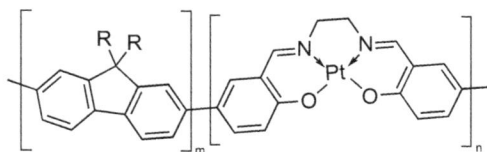

Figure 5.18 A coordination polymer in which phosphorescence results from a transition metal. The platinum salen complex introduces phosphorescence into the polymer.

explain the mechanism of photoluminescence by considering a transition metal-containing π-conjugated polymer being excited to the excited singlet state (S_1) by absorption of a photon. The photoexcited polymer can return to the ground state (S_0) through two possible radiative decay pathways: fluorescence (singlet excited state to ground state, $S_1 \rightarrow S_0$) and phosphorescence (triplet excited state to ground state, $T_1 \rightarrow S_0$) (Figure 5.17). For phosphorescence, intersystem crossing (ISC) from the singlet excited state to the triplet excited state is a prerequisite. The rate of ISC strongly depends on the relative positions of singlet and triplet excited states. Transition metals have large spin–orbit coupling which increases the likelihood of the singlet and triplet excited states to mix, ultimately allowing spin-forbidden phosphorescence to occur. Indeed, transition metal-containing conjugated polymers, such as the polyfluorene-*co*-poly(platinum salen complex) (Figure 5.18), exhibit phosphorescence with long emission lifetimes, making them ideal precursors for fabricating high-efficiency light-emitting diodes. The polymer shown in Figure 5.18 exhibits solution and solid state phosphorescence due to the platinum salen complex.[29] Photoluminescence can be measured in the solution or solid state using photoluminescence spectroscopy discussed in Chapter 4.

5.4.4 Redox Activity

Many inorganic and organometallic polymers incorporate redox-active moieties within their framework. These redox-active moieties, which are transition metal complexes in most cases, impart redox activity to these polymers. Redox-active transition metal complexes such as sandwich complexes, metal salen complexes, metal dithiolenes, and polypyridine metal complexes (Figure 5.19) enable the design of redox-active organometallic and coordination polymers with potential applications as chemosensors, electrocatalysts, molecular batteries, antipathogenic agents, and drug delivery vectors (see the discussion in Chapter 6). As a result of the intrinsic variable oxidation states of the transition metals, redox-active polymers containing these metals can undergo multiple electron transfer processes, which are useful for catalysis or other charge transfer processes. The nature of the metal centre and the ligand field influence the product of the redox process. In metal salen complex-containing polymers, for instance, a one-electron redox process may yield an oxidised species in the form of higher valent metal species ($M^{n+1}L$) or ligand-based radical species ($M^nL^•$) depending on the metal centre or ligand field. Metal salen complexes of transition metals and inner transition metals are desirable because of their multiple oxidation states and photoluminescent properties.

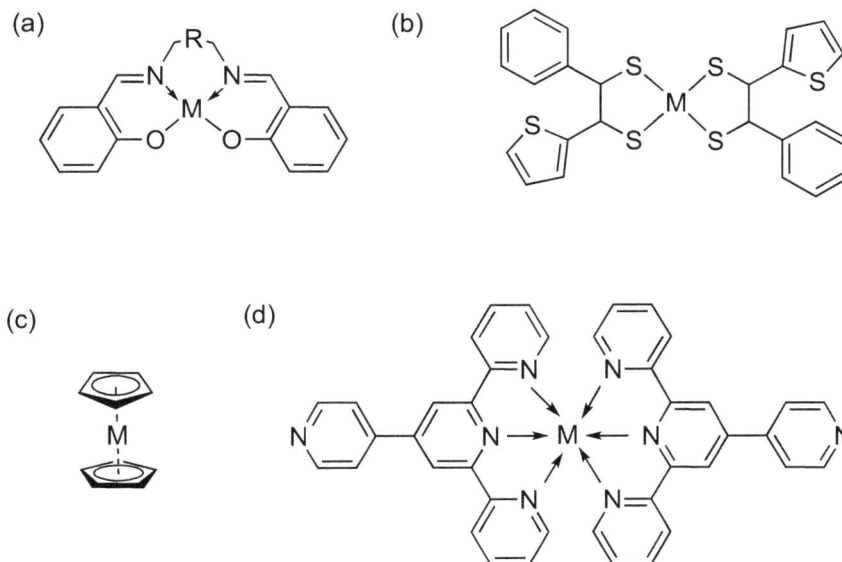

Figure 5.19 Examples of redox-active organometallic complexes used in the design of redox-active polymers: (a) a metal salen complex, (b) a metal dithiolene, (c) a metal sandwich complex, and (d) a metal polypyridine complex.

Box 5.11   Review questions.

1. Some plastics become soft and others stiff when exposed to organic solvents. Provide a likely explanation.
2. Using chemical equations, provide a synthetic route to a redox-active organometallic polymer that self-assembles into spherical micelles.
3. How can you modify the polymer in question 2 to self-assemble into vesicles?
4. Propose a synthesis route to a liquid crystalline inorganic polymer. Describe a structural change that can quench liquid crystallinity.
5. A plastic cup has a T_g of 55 °C. Will this cup be useful as a coffee cup? Propose structural and molecular changes to increase the T_g of the polymer used in fabricating the cup.
6. Explain the effect of high crystallinity on (a) solubility, (b) T_g and T_m, and (c) electrical conductivity.

Another class of redox-active polymers contains sandwich complexes, particularly ferrocene, the archetypal redox-active sandwich compound. Although ferrocene exhibits primarily a one-electron redox process, it is well utilised to synthesise redox-active polymers because of its thermal and oxidative stability and the reversibility of its redox process. In addition, ferrocene reactivity as an electrophile and the number of its derivatives allow access to redox-active polymers of different micro- and macrostructures. Incorporation of other sandwich complexes such as cobaltocene and ruthenocene into the polymeric framework yields redox activity (Box 5.11).

Further Reading

1. A. S. Abd-El-Aziz, C. Agatemor and W.-Y. Wong, *Macromolecules Incorporating Transition Metals: Tackling Global Challenges*, Royal Society of Chemistry Publishing, Cambridge, 2018.
2. J. E. Mark, H. R. Allcock and R. West, *Inorganic Polymers*, Oxford University Press Inc, New York, 2005.
3. R. D. Archer, *Inorganic and Organometallic Polymers*, Wiley-VCH, New York, 2001.
4. G. Odian, *Principles of Polymerization*, John Wiley & Son Inc, Hoboken, 2004.
5. N. P. S. Chauhan and N. S. Chundawat, *Inorganic and Organometallic Polymers*, Walter de Gruyter GmbH, Berlin, 2019.

References

1. S. Abd-El-Aziz, C. Agatemor, N. Etkin and R. Bissessur, *J. Mater. Chem. C*, 2017, **5**, 2268–2281.
2. J. Papanu, D. Hess, D. Soane and A. Bell, *J. Electrochem. Soc.*, 1989, **136**, 3077–3083.
3. M. Tanabe and I. Manners, *J. Am. Chem. Soc.*, 2004, **126**, 11434–11435.
4. U. F. Mayer, J. B. Gilroy, D. O'Hare and I. Manners, *J. Am. Chem. Soc.*, 2009, **131**, 10382–10383.
5. C. N. Jarrett-Wilkins, R. A. Musgrave, R. L. Hailes, R. L. Harniman, C. F. Faul and I. Manners, *Macromolecules*, 2019, **52**, 7289–7300.
6. A. S. Abd-El-Aziz, C. Agatemor, N. Etkin, D. P. Overy, M. Lanteigne, K. McQuillan and R. G. Kerr, *Biomacromolecules*, 2015, **16**, 3694–3703.
7. B. A. Miller-Chou and J. L. Koenig, *Prog. Polym. Sci.*, 2003, **28**, 1223–1270.
8. J. A. Massey, K. Temple, L. Cao, Y. Rharbi, J. Raez, M. A. Winnik and I. Manners, *J. Am. Chem. Soc.*, 2000, **122**, 11577–11584.
9. K. Biradha, Y. Hongo and M. Fujita, *Angew. Chem., Int. Ed.*, 2000, **39**, 3843–3845.
10. D. A. Rider, K. A. Cavicchi, K. N. Power-Billard, T. P. Russell and I. Manners, *Macromolecules*, 2005, **38**, 6931–6938.
11. X.-J. Feng, H.-Y. Han, Y.-H. Wang, L.-L. Li, Y.-G. Li and E.-B. Wang, *CrystEngComm*, 2013, **15**, 7267–7273.
12. G. Cambridge, M. J. Gonzalez-Alvarez, G. Guerin, I. Manners and M. A. Winnik, *Macromolecules*, 2015, **48**, 707–716.
13. J. V. Alsten and S. Granick, *Macromolecules*, 1990, **23**, 4856–4862.
14. T. M. Gädda and W. P. Weber, *J. Polym. Sci., Polym. Chem.*, 2006, **44**, 3629–3639.
15. J. B. Gilroy, S. K. Patra, J. M. Mitchels, M. A. Winnik and I. Manners, *Angew. Chem., Int. Ed.*, 2011, **50**, 5851–5855.
16. M. Cazacu, A. Vlad, M. Marcu, C. Racles, A. Airinei and G. Munteanu, *Macromolecules*, 2006, **39**, 3786–3793.
17. M. K. Lee and D. J. Meier, *Polymer*, 1993, **34**, 4882–4892.
18. A. S. Abd-El-Aziz, C. Agatemor, N. Etkin and R. Bissessur, *Macromol. Chem. Phys.*, 2015, **216**, 369–379.
19. P.-J. Flory, *Proc. R. Soc. London, Ser. A*, 1956, **234**, 60–73.
20. U. Caruso, A. Roviello and A. Sirigu, *Macromolecules*, 1991, **24**, 2606–2609.
21. M. Altmann and U. H. Bunz, *Angew. Chem., Int. Ed.*, 1995, **34**, 569–571.
22. H. R. Allcock and C. Kim, *Macromolecules*, 1989, **22**, 2596–2602.
23. J. Elbert, M. Gallei, C. Rüttiger, A. Brunsen, H. Didzoleit, B. Stühn and M. Rehahn, *Organometallics*, 2013, **32**, 5873–5878.
24. A. K. Andrianov, A. Marin and J. Chen, *Biomacromolecules*, 2006, **7**, 394–399.
25. A. S. Abd-El-Aziz, C. Agatemor and N. Etkin, *Macromol. Rapid Commun.*, 2014, **35**, 513–559.
26. B. Z. Momeni and S. Heydari, *Polyhedron*, 2015, **97**, 94–102.
27. A. J. Heeger, *Rev. Mod. Phys.*, 2001, **73**, 681.
28. H. Nishihara, M. Kurashina and M. Murata, *Macromol. Symp.*, 2003, **196**, 27–38.
29. F. Galbrecht, X. H. Yang, B. S. Nehls, D. Neher, T. Farrell and U. Scherf, *Chem. Commun.*, 2005, 2378–2380.

6 Examples of Inorganic and Organometallic Polymers

6.1 Introduction

The previous chapters discussed the prospects, types, syntheses, characterisation, and properties of inorganic and organometallic polymers. A pertinent question at this point is: are inorganic and organometallic polymers esoteric materials that satisfy the scientific curiosity of a few or materials with real or potential applications? The answer to both questions is yes! Compared with organic polymers, which constitute most of the synthetic polymers on the market, most of the reported inorganic and organometallic polymers are mere laboratory curiosities due to technical and economic issues that include stability and cost. However, a few inorganic and organometallic polymers are now commercial successes due to their unique properties that drive their applications in various fields (see Chapter 7). The prospects for inorganic and organometallic polymers to advance the development of functional materials are promising, given the ubiquity and unique properties of inorganic elements and transition metals (Box 6.2). Indeed, these elements and metals constitute the majority in the periodic table, with most exhibiting chemistries compatible with polymerisation conditions. Many main group elements, such as silicon, phosphorus, and nitrogen, have been successfully incorporated into commercial polymers. Further, among the transition metals, several, including those of the inner transition metals, are moieties in many well-defined and characterised polymers. However, commercial impact in this class is limited at present. For simplicity, we will discuss polymers containing main group elements by subgrouping based on the element's group in the periodic table. At the same time, those containing transition metals will be subgrouped as organometallic and coordination polymers (Box 6.1).

6.2 Polymers Containing Main Group Elements

Several main group elements have been incorporated into organic and inorganic polymers (Figure 6.2).[11-22] These elements include those in group 13, such as boron (B),

Fundamentals of Inorganic and Organometallic Polymer Science
By Christian Agatemor, Kajal Ghosal, Samuel Fura and Peter J. S. Foot
© Christian Agatemor, Kajal Ghosal, Samuel Fura and Peter J. S. Foot 2024
Published by the Royal Society of Chemistry, www.rsc.org

Box 6.1 Learning outcomes.

By the end of this chapter, the student should be able to:

1. Describe and identify inorganic polymers.
2. Describe and identify coordination and organometallic polymers.
3. Describe and identify metal–organic frameworks.
4. Design inorganic or organometallic polymers with a target property.
5. Describe some important properties of polysiloxanes.
6. Describe some important properties of polyphosphazenes.
7. Describe some important properties of MOFs.
8. Give examples of inorganic polymeric glasses.

Box 6.2 Periodic table and polymers.

The periodic table of elements uses the atomic number to arrange elements in rows (period) and columns (group). Since Dmitri Mendeleev compiled the first version in 1869,[1] the table has evolved from 67 to 118 elements with the majority being inorganic elements, transition metals, and inner transition metals. It now provides a quick reference for information on elements, their atomic masses, and chemical symbols. Today, chemists use the table to discern the relationship between the atomic number and properties such as electronegativity, ionisation energy, and chemical reactivity. Generally, elements in the same group exhibit similar chemical reactivity, with non-metallic properties increasing from left to right across the period and metallic properties increasing in the opposite direction.[2,3] This trend enables a qualitative prediction of the chemistry of the elements and builds on the similarities in the electronic configuration of elements within a group or a period. Indeed, chemists can use the table to qualitatively categorise the chemical reactivity and properties of chemical compounds based on elemental composition. Notably, the table provides a quick reference to select elements for designing families of molecules exhibiting similar properties. Many polymer chemists rely on the periodic table to select elements to design families of polymers with similar properties. For example, the discovery of ferrocene,[4,5] the archetypal organometallic compound characterised by an Fe(II) ion sandwiched between two cyclopentadienyl rings (Figure 6.1), inspired the design of congeners by exploring close neighbors of iron with similar electronic configurations in group 8 and period 4. Undeniably, ferrocene

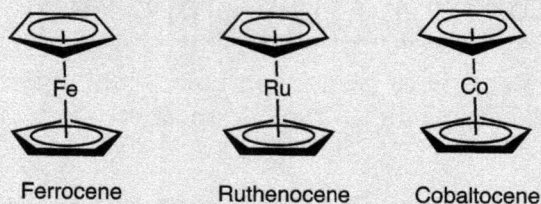

Fe	Ru	Co
Ferrocene	Ruthenocene	Cobaltocene

Figure 6.1 The discovery of ferrocene (bis(cyclopentadienyl)iron(II)) inspires the design of ruthenocene (bis(cyclopentadienyl)ruthenium(II)) and cobaltocene (bis(cyclopentadienyl)cobalt(II)) based on the relationship of iron (Fe) with ruthenium (Ru) and cobalt (Co) in the periodic table.

contributed to the design of ruthenocene and cobaltocene (Figure 6.1).[6,7] Similarly, developing ferrocene-containing polymers nurtured the development of ruthenocene- and cobaltocene-containing polymers.[8–10] For beginners and experts the periodic table is an excellent reference to select elements for designing new inorganic or organometallic polymers or congeners of existing ones.

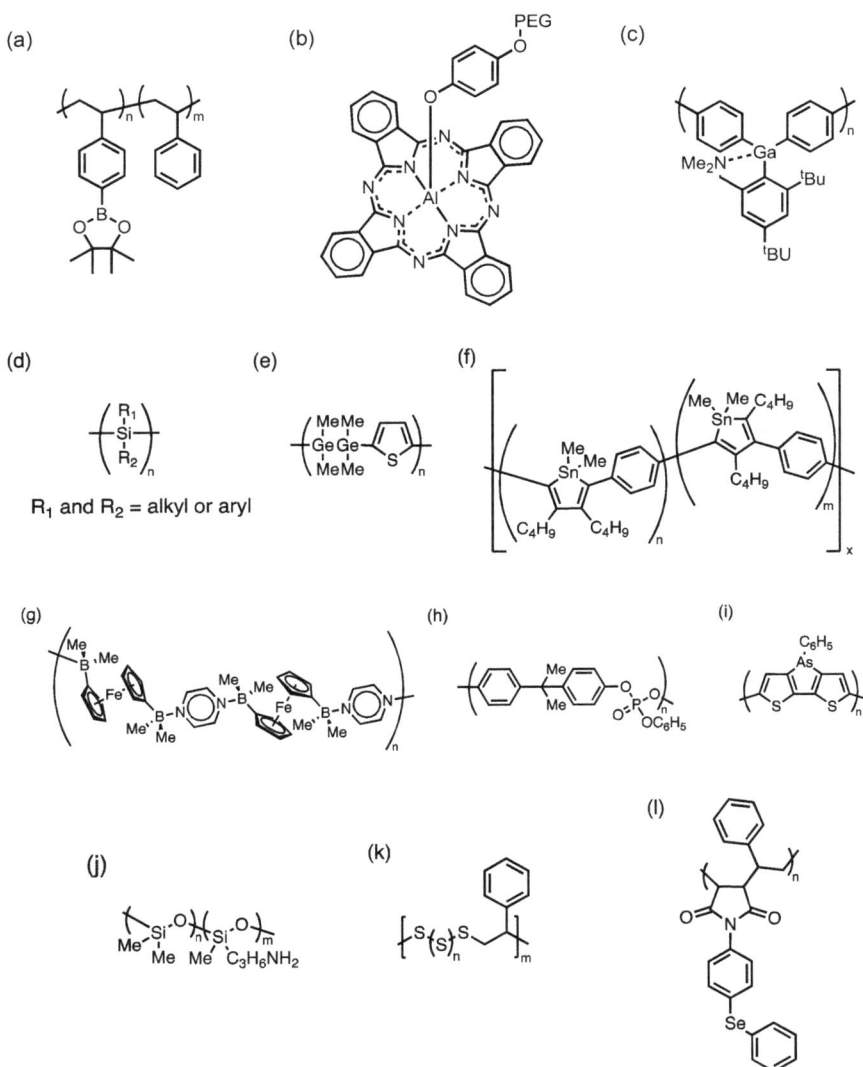

Figure 6.2 Several main group elements, including B, Al, Ga, Si, Ge, Sn, N, P, As, O, S, and Se, have been incorporated into organic, inorganic, and organometallic polymers.

aluminum (Al), and gallium (Ga); group 14, such as silicon (Si), germanium (Ge), and tin (Sn); group 15 such as nitrogen (N), phosphorus (P), and arsenic (As); and group 16 such as oxygen (O), sulfur (S) and selenium (Se) (Figure 6.2). Some of these polymers, such as polysiloxanes (Figure 6.2j), are now commercially available, while others, such as arsenic-containing polymers (Figure 6.2i), are yet to enter the market. We will discuss only polymers in which a covalent bond connects the main group elements in the polymer backbone, taking examples from the p-block elements, except for the group 17 (halogens) and 18 (noble gases) elements. We will refrain from discussing the polymers incorporating group 18 since none apparently exists at present. Indeed, the group 18 elements (note: helium is an s-block element but is positioned in group 18 due to its chemical similarity to elements in this group) have full valence electron shells, making them highly stable and less likely to participate in chemical reactions, including polymerisation. We will also avoid discussing polymers incorporating group 17 elements, notwithstanding the presence of halogens in various polymers or macromolecules. The halogens, unlike the noble gases, are highly reactive, forming polyhalogen ions, such as polyiodide ions $[I_{29}]^{3-}$, or organohalides such as tetrafluoroethylene, the monomer for synthesising polytetrafluoroethylene (Teflon), a prime example of halogenated polymers. Nonetheless, we omit polymers containing the group 17 elements because the formation of polyhalogen ions is better described by catenation[23,24] (monomer molecules are not involved in the process of forming the polyhalogen ion), not polymerisation (Box 6.3), and halogenated polymers are better classified as organic polymers (see Chapter 1). Further, polymers such as sodium or calcium alginate containing s-block elements (groups 1 and 2 elements) presently exist in the market. Because group 1 and 2 elements are countercations in these polymers, we will not use them as examples because they contribute nothing to the polymer backbone structure. It is worth mentioning, however, that some coordination polymers, including metal–organic frameworks (MOFs) (discussed in Section 6.3.1), incorporate some group 1 and 2 elements within their framework.[25] Still, these coordination polymers are yet to make it to the market. In this section, we will use polymers containing group 13 to 16 elements as examples because these

---

Box 6.3   Polymerisation *versus* catenation.

- Polymerisation is bonding monomers or a mixture of monomers to form a polymer (see Chapter 1). The constitutional unit in a polymer is derived from the monomer molecule.
- Catenation is bonding atoms of the same element to form chains or rings. The ability of an element to catenate may contribute to polymerisation. For example (Figure 6.3), sulfur (S_8) exists as a puckered or crown ring of eight atoms, the orthorhombic polymorph, formed through catenation. On heating, this ring opens and links together *via* catenation, yielding a polymer made from several S_8 monomers. The ability of orthorhombic sulfur to polymerise is linked to catenation, the ability of sulfur atoms to self-link through a covalent bond.

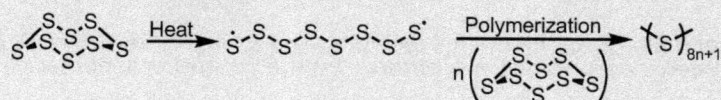

Figure 6.3   Catenation results in cyclo-octasulfur, which undergoes ring opening upon heating and undergoes polymerisation due to the ability of sulfur to catenate.

elements can form a covalent bond and contribute to the overall structure and function of the polymer.

6.2.1 Polymers Containing Group 13 Elements

The group 13 elements include a non-metallic element, boron (B), and four metallic elements, aluminum (Al), gallium (Ga), indium (In), and thallium (Th). These elements, specifically boron and aluminum, are constituents of many materials. For instance, borax ($Na_2[B_4O_5(OH)_4] \cdot 8H_2O$), a compound containing boron, has been used in pottery glaze since ancient times, while kaolin ($Al_2Si_2O_5(OH)_4$), a clay mineral containing aluminum, is the oldest known ceramic material in history. Therefore, it is rational to argue that using B and Al in materials engineering predates modern polymer science and engineering. The most unusual of the group 13 elements is boron, which is more like its horizontal neighbor, carbon (C), and diagonal neighbor, silicon (Si), than the other group 13 elements. Like the hydrocarbons of carbon, boron bonds with hydrogens to form hydrides called boranes; like the silicates of silicon, it bonds with oxygen to form borates. Notably, in boranes, hydrogen bridges two boron atoms in a nonclassical three-centre two-electron bond. We will focus on boron-containing polymers because of the unique chemistry of boron relative to the other group 13 elements and the potential applications of these polymers in materials engineering (see Chapter 7).

6.2.1.1 Polyborazylene

Several parallels exist between carbon and boron compounds to motivate the exploration of boron in inorganic polymer synthesis akin to carbon in organic polymer synthesis. One parallel is between the structure and properties of benzene, C_6H_6, and the isoelectronic inorganic ring, borazine, $B_3N_3H_6$ (Figure 6.4).[2] This similarity suggests the feasibility of synthesising polyborazylene, the borazine analog of polyphenylene (Figure 6.4). Indeed, polyborazylene, derived by coupling borazine rings through a boron–boron bond, a boron–nitrogen bond, or a bridging atom/molecule, is synthetically accessible. Early works explored metal-catalysed coupling, but later developments demonstrated that controlled thermal dehydropolymerisation of borazine *in vacuo* at moderate temperature produces polyborazylene in excellent yield.[26–28]

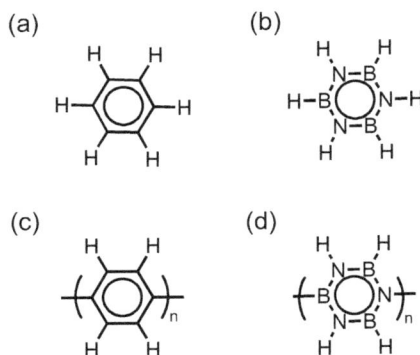

Figure 6.4 (a) Benzene, (b) borazine, (c) polyphenylene, and (d) polyborazylene.

(a)

(b)

(c)

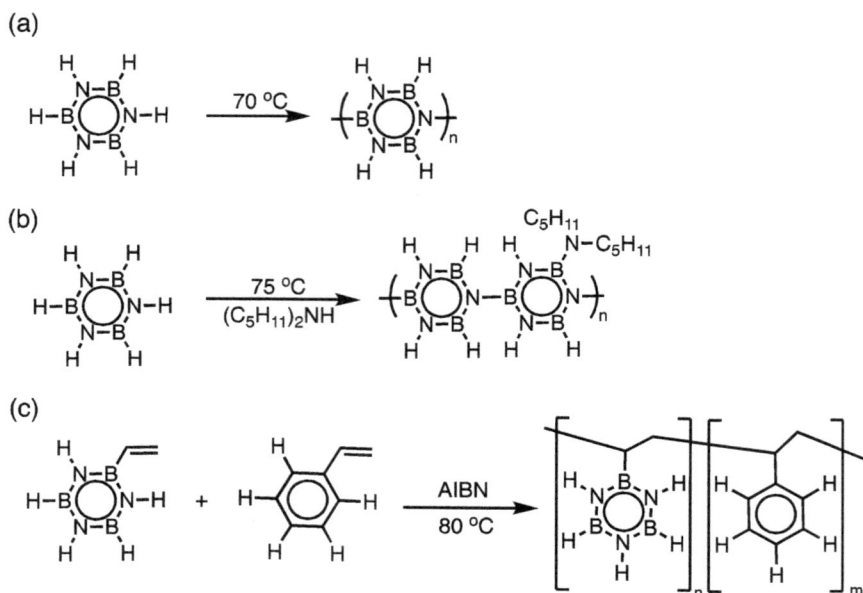

Figure 6.5   (a) Thermal dehydropolymerisation of borazine to polyborazylene. (b) Synthesis of amino-modified polyborazylene. (c) Synthesis of poly(styrene-co-B-vinylborazine).

A typical dehydropolymerisation involves heating borazine *in vacuo* at 70 °C for approximately 48 hours with periodic degassing to make a viscous liquid, which eventually forms a white solid polyborazylene after evaporation *in vacuo* (Figure 6.5).[26] Solid polyborazylene is highly soluble in polar solvents such as tetrahydrofuran and glycol ethers, enabling detailed analytical, spectroscopic, and molecular weight characterisations. From the elemental analysis of the purified polyborazylene, an empirical formula of $B_{3.1}N_{3.0}H_{2.7}$ was found, which deviates from the expected $B_3N_3H_4$, and suggests a branched chain or partially crosslinked structure.[26] Data from low-angle laser light scattering and UV absorbance further support a highly branched or partially cross-linked polymer. The molecular weights depend on the duration of polymerisation, with a duration of 48 hours yielding polymers with $M_w = 7600 \pm 460$ Da and $M_n = 3400 \pm 210$ Da, while 24 hours affording polymers with $M_w = 2100 \pm 330$ Da and $M_n = 980 \pm 150$ Da.[26] As a solid, polyborazylene is stable for an extended time at room temperature *in vacuo* but decomposes over several hours when exposed to moist air. The tendency of polyborazylene to crosslink at low temperatures limits its applications. Strategies to minimise crosslinking include modifying the borazine ring with substituents to reduce the number of reaction sites.[28] Also, borazine-containing copolymers composed of borazine moieties and other polymers allow access to materials with unusual or enhanced properties. Copolymerisation of B-vinylborazine with other monomers such as styrene yields poly(styrene-co-B-vinylborazine) (Figure 6.5).[27]

6.2.1.2   Polyborates

Polymeric borate glasses are formed from the reaction of boric acid with metal oxides. In boric acid, boron is electron deficient, readily accepting a pair of electrons from a metal oxide ion to form a borate anion. These borate anions are typically boron oxyanions, such as orthoborate ($[BO_3]^{3-}$), metaborate ($[B_3O_6]^{3-}$), diborate ($[B_2O_5]^{4-}$), and tetraborate ($[B_4O_7]^{2-}$), with structures ranging from two- and three- to four-connected polymer networks. Unlike the fixed coordination geometry of central cations in other classes of inorganic oxyanions such as silicates, phosphates, nitrates, vanadates, and carbonates, the overlap of boron 2s and 2p orbitals can result in sp, sp^2, or sp^3 hybridised orbitals, creating a two-connected linear, three-connected trigonal planar or four-connected tetrahedral geometry at the boron atom, respectively.[29] The possibility of multiple coordination geometries coupled with the rich library of counter-cations derived from main group metals, transition metal complexes, or organic ligands results in a structurally diverse family of polyborates.[30–32] Polymeric metal borate glasses are typically network polymers composed of mixed trigonal planar and tetrahedral boron atoms (Figure 6.6). An excellent example of polymeric metal borate glasses is borosilicate glass, typified by Pyrex® products used for many laboratory glassware and kitchenware (Figure 6.6). Borosilicate glasses are produced by fusing boric oxide (B_2O_3), molten silica (SiO_2), soda ash, and alumina to form a network polymer where boron replaces some silicon atoms in the silicate. Similarly, borophosphate glasses, four-connected network polymers, are formed by fusing phosphoric acid, boric acid, and appropriate amounts of alkali metal carbonates, alkaline earth metal oxides, or carbonates.

6.2.1.3 Polymers Containing Other Group 13 Elements

We have discussed polyborazylenes and polyborates as classic examples of inorganic polymers containing group 13 elements because of their potential applications in materials engineering. Polyborazylene, for instance, is a precursor for boron nitride and carbonitride ceramics, whereas polyborate is a precursor for borosilicate glasses, as mentioned above. Other boron-containing polymers, such as borane polymers, have been investigated in the laboratory.[33–37] Metal-catalysed dehydrocoupling of alkylaminoborane or alkylphosphinoborane yields high molecular weight polyaminoboranes or poly(alkylphosphinoboranes) (Figure 6.7), boron–nitrogen analogs of polyolefins.[34,38] Polymers containing other group 13 elements, such as aluminum and gallium, have also been synthesised and investigated for their properties. An example is aluminosilicate glasses, a congener of borosilicate glasses, produced by fusing alumina (Al_2O_3) with silica and other metal oxides. Additionally, the group 13 elements are used in synthesising inorganic polymers and employed in the design of organometallic and coordination polymers (Figure 6.7).[11,39–46] Of particular interest is the use of group 13 elements,

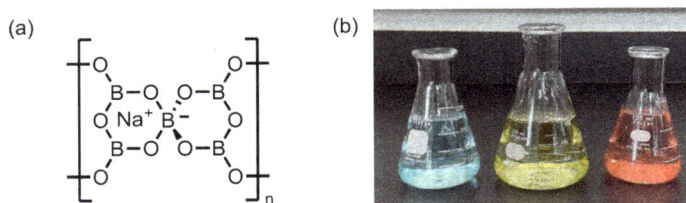

Figure 6.6 (a) Polyborates are typically composed of trigonal planar and tetrahedral boron atoms. (b) Pyrex® glassware used in chemistry laboratories are made from borosilicate glasses (picture courtesy of Christian Agatemor).

specifically aluminum, gallium, and indium, in synthesising MOFs, coordination polymers assembled from organic and inorganic building blocks.

6.2.2 Polymers Containing Group 14 Elements

The group 14 elements include carbon (C), a nonmetal, silicon (Si) and germanium (Ge), semiconductors, and tin (Sn) and lead (Pb), metals. The +4 oxidation state dominates the chemistry of these elements, but the +2 oxidation state becomes dominant at the bottom of the group. Carbon, an element essential to life, defines organic polymers and contributes to the framework of many organometallic and inorganic polymers. Catenation (Box 6.3) is especially pronounced in carbon chemistry, accounting for the ubiquity of carbon-based compounds, including numerous organic polymers. However, catenation is less prevalent in the chemistry of the other group 14 elements. Inorganic polymers containing Si, Ge, Sn, and Pb have been synthesised with silicate glasses, such as borosilicate glasses (Section 6.2.1), and polysiloxanes being exemplary commercially successful inorganic polymers. Because of the commercial success of silicate glasses and polysiloxanes, we will focus on only Si-containing inorganic polymers to exemplify polymers incorporating the group 14 elements (Box 6.4).

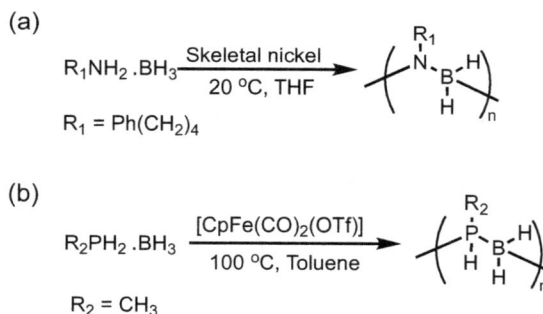

Figure 6.7 Metal-catalysed dehydrocoupling of (a) alkylamineborane and (b) alkylphosphinoborane.

Box 6.4 Practice question.

Question

The hydrides of carbon, such as methane and CH_4, are known for their stability, whereas those of the other elements, such as silane (SiH_4), germane (GeH_4), stannane (SnH_4) and plumbane (PbH_4), are less stable. Explain.

Answer

Silicon, germanium, tin, and lead are less electronegative than hydrogen. Therefore, in silane, germane, stannane, and plumbane, the Si, Ge, Sn, and Pb atoms are electrophilic and more susceptible to nucleophilic attack (carbon is more electrophilic than hydrogen).

6.2.2.1 Silicate Glasses

We discussed borosilicate (Section 6.2.1) and aluminosilicate glasses as examples of boron- and aluminum-containing inorganic polymers. These silicate glasses are also examples of silicon-containing inorganic polymers. Since our previous discussion of silicate glasses focused on synthesis, we will discuss other aspects, specifically the structure and properties. Historically, silicate glasses are among the oldest manmade materials, with glassmaking predating iron smelting. Silicon-containing glasses also exist in nature. Classic examples include obsidian (Figure 6.8), a volcanic glass rich in silica and produced from a rapidly cooled felsic lava extruding from a volcano. Generally, glass or vitreous solid refers to materials formed from a rapidly cooled melt; silicate glass being any vitreous material with a high content of silica. The common structural unit of silicon-based glasses is the tetrahedral SiO_4 building block. Produced by melting and rapidly cooling silica with other materials, silicate glasses differ in composition and properties from a fused-silica glass made from pure silica. Fused-silica glass is a chemically inert glass characterised by a high glass transition temperature that enables applications under harsh chemical and high temperature conditions. A caveat, however, is that fused-silica is susceptible to reaction with hydrofluoric acid and strong alkaline solutions, limiting applications under these conditions. Also, the relatively high cost of fused silica restricts applications to the specialty materials such as optical materials used in fibre optics.

Structurally, the building block of silicate glasses is tetrahedral SiO_4 (Figure 6.8), which shares oxygen atoms to form linear chains, cyclic motifs, layered structures, or three-dimensional networks. The additional cations in silicates exhibit various coordination numbers, ranging from 4 for Be^{2+} to 8 for Ca^{2+}, relative to the oxide anion (O^{2-}). The most common silicate glass is the soda–lime glass, formed by adding soda (sodium carbonate, Na_2CO_3) to lower the glass transition temperature of silica. Because soda is water soluble, lime (calcium oxide, CaO), magnesium oxide (MgO), and aluminum oxide (Al_2O_3) are added to enhance chemical durability, especially water resistance. Properties, such as hardness and softening temperature, can be tuned by varying the amount of the different metal oxides. For example, optically clear but harder glasses are obtained by reducing the amount of sodium and increasing the amount of potassium in soda–lime glasses. Incorporating a transition metal oxide imparts color to soda–lime glasses. Adding cobalt oxide (CoO) or copper(II) oxide (CuO) transforms a characteristically colorless soda–lime glass to deep blue or turquoise colored glass, respectively. Further, adding germanium oxide to form germanosilicate glass increases the refractive index

(a)   (b)

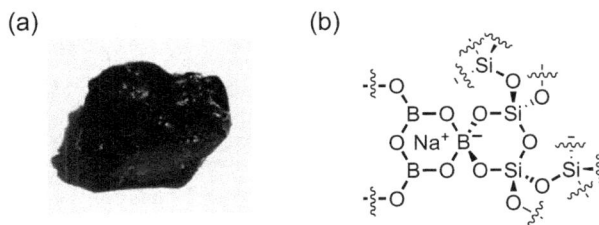

Figure 6.8 (a) Obsidian, a volcanic glass, rich in silica (obtained from pixabay.com). (b) Structure of sodium borosilicate glass.

relative to fused-silica glass. The enhanced refractive index is a desirable property for materials used for making optical fibres. Adding phosphorus pentoxide to silicate glasses results in phosphosilicate glasses, a potentially useful material for fibre optics when Raman frequency shift relative to silica-based fibres is desirable.

6.2.2.2 Polysiloxanes

Polysiloxanes, commonly known as silicones, are inorganic polymers composed of silicon–oxygen backbones and pendent organic groups on the silicon atoms (Figure 6.9). The long Si–O bond length (1.64 Å *versus* 1.54 Å for the C–C bond) and wide Si–O–Si bond angle (143° *versus* 109.5° for the C–C–C bond angle) increase chain flexibility even at low temperatures.[47] Further, the high Si–O bond strength (450 kJ mol^{-1} *versus* 350 kJ mol^{-1} for C–C) enhances stability in oxidative, chemical, biological, and high-temperature environments.[47] Consequently, polysiloxanes feature improved chemical and biological stability and high thermal stability over a wide temperature window (-100 to 250 °C). This exceptional combination of properties, especially chemical and biological stability, drives their wide-spread use in biomedical applications. Typical of all polymers, the properties of poly-siloxanes can be tuned by changing the structure of the pendent group, polymer molecular weight, or polymer architecture. For example, polydiphenylsiloxane is a rigid, crystalline polymer with high melting temperatures, while polydimethylsiloxane is a very flexible, amorphous polymer with very low glass transition temperatures (Figure 6.9).[48] In addition to the thermal properties, bulk properties such as viscoelasticity can be tuned by varying the structure or number of pendent groups. For example, increasing the number of phenyl groups per silicon atom slows the increase in the elastic modulus, while substituting the

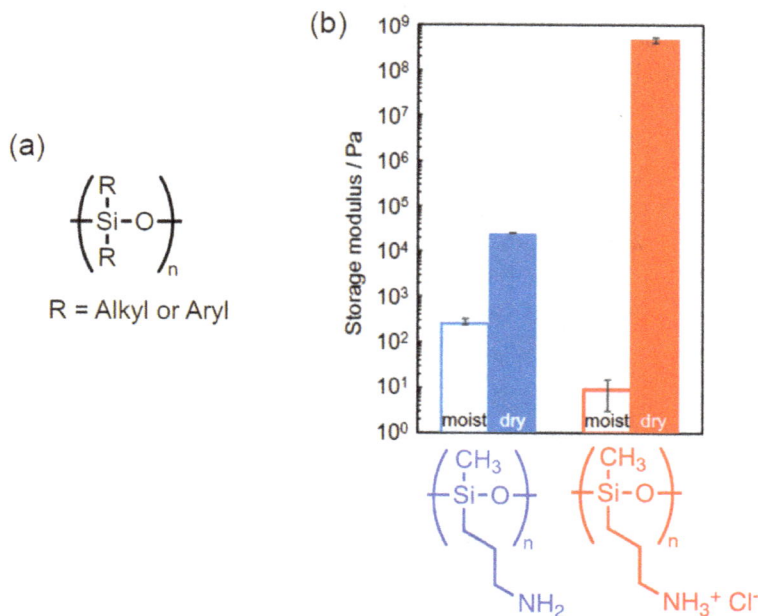

Figure 6.9 (a) Structure of polysiloxanes. (b) Converting the amino groups in poly(3-amino-propylmethylsiloxane) to ammonium hydrochloride enhances the storage modulus. Adapted with permission from ref. 51.

phenyl group with the methyl group has the opposite effect.[49] The impact of the structure on the elastic modulus occurs primarily through molecular weight changes.

The global polysiloxane market was about USD 16.7 billion in 2021 and is projected to reach USD 23.4 billion by 2026. The projected increase is informed by increasing demand for polysiloxane elastomers, fluids, resins, or gels in many industries ranging from construction, personal care, healthcare, and electronics to automobile. This increase in demand is due to the attractive properties such as flame retardance, water repellency, high thermal stability, chemical and biological inertness, good electrical insulation, low surface tension, and high gas permeability of polysiloxanes. Elastomers dominate the polysiloxane market and differ from organic polymer elastomers in the way their mechanical properties respond to reinforcing fillers or changes in the polymer structure. For example, chemical crosslinking coupled with hydrogen bond-induced dynamic crosslinking improves the tensile strength, Young's modulus, and toughness of poly-dimethylsiloxane.[50] Also, chemically converting the amino group in poly(3-aminopropylmethylsiloxane) to ammonium hydrochloride significantly enhances the elastic modulus (Figure 6.9).[51] The presence of water significantly alters the modulus, which changes by a factor of 100 million between a moist and a dry atmosphere. In the dry state, linear poly(3-aminopropylmethylsiloxane) hydrochloride has a shear strength of about 1.5 MPa, functioning as a good adhesive for glass substrates to suspend a 6 kg weight.[51] The attractive properties of polysiloxane elastomers are expanding their applications to different industries as sealants, adhesives, gaskets, electrical insulators, coatings, prosthetic devices, contact lenses, and artificial heart valves, among others.

Polysiloxane fluids are mainly linear polysiloxanes obtained by the hydrolysis of dichlorosilanes in the presence of excess water. The polymerisation yields a mixture of cyclic oligomers and linear polysiloxanes, which can be used as silicone fluids. Resins are prepared through hydrolysis but include trichlorosilanes alongside dichlorosilanes to introduce crosslinking and increase molecular weights. Polysiloxane elastomers can be obtained by ambient or high-temperature crosslinking of oligomers or monomers and are characterised by a degree of polymerisation as high as 11 000.[47] This crosslinking is achieved with silane monomers or polysiloxane oligomers modified with reactive hydroxy, alkoxy or vinyl end groups that can undergo further reaction to increase the degree of polymerisation or crosslinking in the polymer. Depending on the reactive end group, the crosslinking is catalysed by a basic catalyst, such as zinc octanoate; organometallic catalysts, such as dibutyltin laurate; transition metal catalysts, such as platinum catalysts; or organic peroxides, such as benzoyl peroxide.[47]

6.2.2.3 Polysilanes

Polysilanes are silicon-containing polymers composed of a backbone of catenated silicon atoms bonded through σ bonds and attached to hydrogen atoms or organic groups (Figure 6.10). As expected, the structure of the pendent group dramatically influences the properties of the polymer, especially crystallinity. If the pendent groups on the silicon atom are identical to those in polydimethylsilane or polydiphenylsilane, highly crystalline polymers can be obtained, whereas if they differ, the symmetry decreases, lowering crystallinity. These polymers exhibit many interesting properties, including unusual optical and electronic properties due to the delocalisation of the

(a)   (b)   (c)

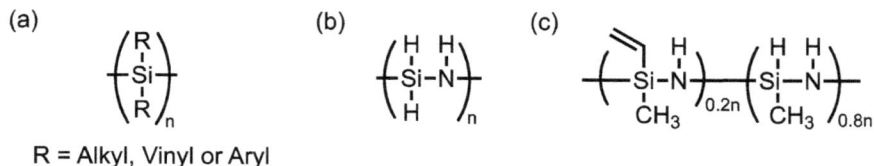

R = Alkyl, Vinyl or Aryl

Figure 6.10 (a) Structure of (a) a polysilane, (b) a perhydropolysilazane, and (c) Durazane 1800.

σ-bond electrons along the silicon backbone. The delocalisation may stem from the interaction between the vacant d-orbitals of the relatively large adjacent silicon atoms. Indeed, evidence from ultraviolet spectroscopy shows that polysilanes exhibit strong ultraviolet absorption at 300–400 nm[52,53] which may be due to the low ionisation energy of the Si–Si σ bond that enables σ–σ* electronic transitions. This photoactivity can result in photolytic cleavage, precluding some light-related applications of the polymers but opening new opportunities in positive photoresist fabrication.

As discussed in Chapter 2, the modified Wurtz coupling of dichlorodiorganosilanes (RR′Cl$_2$Si; R and R′ are organic groups) is currently the primary synthesis method for high molecular weight, linear polysilanes.[54] However, a trichloromonoorganosilane (RCl$_3$Si) yields a three-dimensional crosslinked polyalkylsilyne (RSi)$_n$.[55] Dehydrogenative coupling of primary organosilanes catalysed by transition metal complexes such as Cp$_2$ZrR$_2$ (Cp = cyclopentadienyl ligand, R = hydrogen or alkyl group) is an alternative synthesis route to low molecular weight atactic linear polysilanes.[56] Notwithstanding their optical and electronic properties, the current interest in polysilanes stems from their use as pre-ceramic polymers (see Chapter 7). Thermolysis of polysilanes in an inert atmosphere at 450 °C transforms them into polycarbosilanes, which can be spun into fibres and pyrolysed at 1200 °C to form a β-silicon carbide ceramic (see Chapter 7). Nevertheless, the optical and electronic properties of polysilanes make them potentially useful as semiconductors, photoconductors, and nonlinear optical materials (see Chapter 7).

6.2.2.4 Polysilazanes

Polysilazanes are silicon-containing polymers with a backbone composed of a silicon–nitrogen bond, a pendent hydrogen atom, and alkyl and/or aryl groups on the silicon atom (Figure 6.10). If the pendent groups are hydrogen atoms, the polymer is termed a perhydropolysilazane or an inorganic polysilazane, and when it is an organic group, it is called an organopolysilazane (Figure 6.10). On pyrolysis above 400 °C, polysilazanes transform to Si$_3$N$_4$, SiON, SiCN, SiCNO, and SiC ceramics depending on the pyrolysis conditions and chemical composition of the polymer. Compared with most polymers, polysilazanes feature higher hardness, enhanced thermal and chemical stability, and excellent scratch, abrasion, impact, and weathering resistance. Polysilazanes are good coating materials because of the reactivity of the Si–H or Si–N–Si bond towards hydroxyl groups on the surface of substrates such as plastics, metals, or ceramics to form a silicon–oxygen bond (see Chapter 7).[57–59] These coatings can tune the properties of the substrate surface, imparting improved mechanical and chemical resistance or adjusting the thermal conductivity, biocompatibility, wettability, permeability, and photoactivity to meet the needs of a specific application. Also, some polysilazanes have excellent anti-adhesion and easy-to-clean properties comparable to poly(tetrafluoroethylene)-based coatings. Commercially available as TPdur Bright® (durXtreme GmbH, Germany) or Durazane® (Merck KGaA, Germany), these polymers are mainly used in high-performance protective coatings in premium-value applications to ensure high temperature stability, water, and dirt repellency, scratch protection and corrosion prevention. Some commercially available polysilazanes, such as the perhydropolysilazane Durazane 2250 (Merck KGaA, Germany), are liquids, while others, such as the organosilazane Durazane 1800 (Merck KGaA, Germany), are soluble solids (Figure 6.10). The ability to constitute these polymers as liquids or solutions is advantageous in coating applications, as it allows liquid phase deposition ranging from simple dipping and spinning to spraying. After deposition, crosslinking of polysilazanes can be used to enhance the thermal, mechanical, and chemical stability of the coating.[57]

6.2.3 Polymers Containing Group 15 Elements

The group 15 elements include two nonmetals (nitrogen and phosphorus), two metalloids (arsenic and antimony), one metal (bismuth), and one uncharacterised radioactive element (moscovium). A defining characteristic of these elements is that they have five valence electrons in the ground state and a tendency to form covalent double or triple bonds. Among these elements, nitrogen is the most important from the biological and chemical perspectives. Nitrogen occurs in all living organisms as a component of proteins, nucleic acids, and in many essential chemicals, including ammonia, nitric acid, nitrate, and cyanides. Similarly, phosphorus is vital to life, being present in nucleic acids and hydroxyapatite, the main component of bone. The other members of the group are less relevant to life and less used in the chemical industry. Arsenic salts are highly toxic but arsenic, like phosphorus, is an n-type dopant in semiconductors. However, antimony and bismuth compounds are less toxic than arsenic, with bismuth subcarbonate ((BiO_2)CO_3) finding applications in treating gastrointestinal disorders, including peptic ulcers. Bismuth is also increasingly used to make glasses, ceramics, and alloys, replacing lead in lead-free solders.[2] All the group 15 elements, except moscovium, have been incorporated into inorganic and organometallic polymers. However, only nitrogen and phosphorus are the constituents of natural biopolymers, such as proteins and nucleic acids, discovered to date. Nitrogen and phosphorus are the defining elements in polyphosphazenes, a class of inorganic polymers. Some organic polymers, such as polyacrylamides, containing nitrogen but composed of carbon–carbon backbones are excluded from this discussion. This section will focus on polyphosphazenes and polyphosphates as examples of polymers containing group 5 elements. Polyborazylenes (discussed in Section 6.2.1) also exemplify polymers with a group 13 element, specifically nitrogen, in the backbone.

6.2.3.1 Polyphosphazenes

A distinctive feature of phosphazenes is the presence of a phosphorus(v) atom conjugated to a nitrogen(iii) atom through a double bond (Figure 6.11). Polyphosphazenes are typically composed of an inorganic backbone of alternating phosphorus(v) and nitrogen(iii) atoms and pendent groups attached to the phosphorus atom. The pendent group may be inorganic, resulting in a wholly inorganic polymer or an organic polymer,

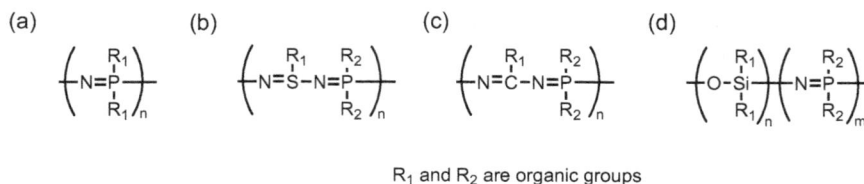

(a)   (b)   (c)   (d)

R₁ and R₂ are organic groups

Figure 6.11 Structure of a (a) polyphosphazene, (b) polythiophosphazene, (c) polycarbopho-sphazene, and (d) poly(phosphazene-*block*-siloxane).

yielding a hybrid organic–inorganic polymer (Figure 6.11). It is also synthetically feasible to append an organometallic group as a pendent group on the phosphorus atom, replace some phosphorus atoms with other atoms such as carbon and sulfur, or conjugate polyphosphazene to other polymers such as polysiloxanes to form a copolymer (Figure 6.11). Different architectures, including linear, star, dendritic or comb-type, can be obtained synthetically, making polyphosphazenes one of the most structurally diverse and largest classes of inorganic polymers.

Melt ring-opening polymerisation (ROP) (see Chapter 3) of hexachlorocyclotriphosphazene at 220–250 °C is the general synthesis route to high molecular weight polyphosphazenes. Crosslinking and branching occur during the ROP at high conversion, producing insoluble polymers; therefore, limiting conversion to 70% is advisable.[47] The use of a Lewis acid catalyst, such as boron trichloride, or an inert solvent, such as carbon disulfide, allows polymerisation to higher conversions at lower temperatures with the benefit of yielding non-crosslinked soluble polymers.[47] Evidence from electrical conductivity measurements suggests that melt ROP of hexachlorocyclotriphosphazene proceeds *via* a cationic mechanism involving the ionisation of the polar phosphorus–chloride bond.[47] This evidence is further supported by the observation that hexaalkylcyclotriphosphazene does not undergo polymerisation because the less polar phosphorus–carbon bond is not ionisable under these experimental conditions.

Nonetheless, derivatives of polyphosphazenes can be obtained since the chloride in the polymer can readily undergo a nucleophilic substitution reaction and be replaced with a nucleophile. For example, using sodium alkoxides or alkylamines the alkoxyl or amino derivative of a polyphosphazene can be obtained. However, the use of organometallic nucleophiles such as phenyllithium to obtain alkyl or aryl derivatives is inefficient as the phosphorus–nitrogen bond cleaves once the degree of substitution exceeds 10%.[60,61] A more efficient route uses *N*-(trimethylsilyl)-*P,P*-dialkyl-*P*-halophosphoranimine, which undergoes condensation polymerisation that proceeds *via* a living cationic chain mechanism at ambient temperature to yield low molecular weight polymers with narrower molecular weight distributions.[62–64] The living character of the polymerisation allows the synthesis of block copolymers with architecture ranging from linear and star to dendritic. The properties of polyphosphazenes are highly tunable by varying the molecular weight, the polymer's crosslinking density, or the pendent group's structure and functionality. A pendent alkoxyl group imparts water repellency, whereas a fluorinated alkyl group imparts chemical inertness. Indeed, polyphosphazenes with a pendent trifluoroethyl group are sufficiently inert for fabricating artificial blood vessels and organs. Although no commercial application of polyphosphazenes exists, their highly tunable properties suggest potential applications as drug delivery systems, solid electrolytes, and fire retardants.

6.2.3.2 Polyphosphates

These are polymers derived from the repeat units of tetrahedral phosphate groups linked by oxygen atoms. Naturally, polyphosphates are ubiquitous, occurring in various organisms and speculated to play a critical role in prebiotic chemistry as the primordial source of phosphates in nucleic acids.[65,66] Depending on the nature of the counterion, a completely inorganic or hybrid organic–inorganic polymer results. An inorganic polymer results if the counterion is a metal or ammonium ion, while a hybrid polymer is formed when an organic group is conjugated with oxygen (Figure 6.12). Several linear, cyclic, or crosslinked polyphosphates have been reported in the literature. Alkali metal (M)-containing linear polymers have the general formula: $M_{(n+2)}P_nO_{(3n+1)}$, where n, the degree of polymerisation, may vary from 2 to 10^6.[66] Polyphosphates are synthetically accessible by step polymerisation that involves thermal dehydration of the corresponding metal or ammonium dihydrogen phosphate (Figure 6.12). The structure of the counterion or synthesis method and conditions influence the properties, including the molecular weight and crystallinity of the resulting polymer. For example, the glass transition temperature (T_g) of alkali metal polyphosphates is 200 °C lower than that of alkaline earth metal polyphosphates.[67] Also, the T_g increases with decreasing size of the cation.[67] Generally, polyphosphates have many valuable characteristics, including affordability, nontoxicity, biodegradability, cation chelation, and water binding, that contribute to their commercial applications. Polyphosphates are among the few commercialised inorganic polymers. Some commercially available polyphosphates are ammonium polyphosphate, used as a fertiliser, and sodium polyphosphate, used as a food additive.

6.2.4 Polymers Containing Group 16 Elements

Group 16 comprises three nonmetals (oxygen, sulfur, and selenium), a metalloid (tellurium), and two radioactive metals (polonium and livermorium). Apart from the radioactive elements, which pose a safety risk, all the other group 16 elements have been incorporated into synthetic polymers, imparting new structures or functions to meet the needs of specific applications. Also, apart from the radioactive elements, all the group 16 elements exhibit similar chemistries, forming ions having an oxidation state of −2 in the reaction with electropositive metals. Nevertheless, the tendency to form compounds in an oxidation state of −2 decreases down the group, with other oxidation states such as +4 and +6 dominating the chemistry. Indeed, sulfur, selenium, tellurium, and polonium form compounds in a +6 oxidation state, namely, sulfates, selenates, tellurates, and polonates, respectively. Oxygen is the most electronegative of the elements in this group

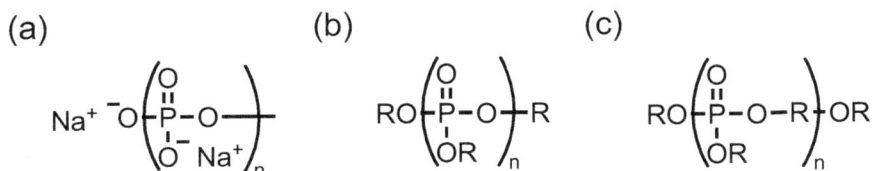

Figure 6.12 Structure of (a) inorganic polyphosphate and (b and c) hybrid polyphosphates.

and a powerful oxidising agent, reacting with most elements at high temperatures. Section 6.2.2 discusses polysiloxanes, a preeminent example of inorganic polymers containing oxygen within their backbones. Sulfur is also a reactive element, capable of reacting with most elements or hydrocarbons during the vulcanisation of rubber. Vulcanisation toughens soft rubber by replacing the hydrogen in C–H bonds with chains of sulfur atoms that crosslink the rubber backbone. Although less reactive, selenium and tellurium exhibit similar chemistries, forming compounds in the −2 oxidation state similar to oxides and sulfides in addition to +2, +4, and +6 oxidation states.

Because we have discussed polysiloxanes, we will focus on polysulfides, polymers containing sulfur atoms. Catenation plays a significant role in sulfur chemistry, accounting for the formation of sulfur rings that can undergo ROP to form homopolymers (Box 6.3). These homopolymers are low molecular weight polymers with inferior technical and commercial relevance compared with hydrocarbon polymers such as polyethylene. Alkylene polysulfides, also known as thiokols, are the only high molecular-weight sulfur polymers with practical applications. Since thiokols are essentially organic polymers due to the presence of the carbon–carbon linkage in the backbone, we will not discuss them in this monograph.

6.2.4.1 Polymeric Sulfur

Sulfur exists in several allotropes, with the cyclo-α-octasulfur (orthorhombic, α-S_8) being the most stable at room temperature. α-Octasulfur is an eight-membered ring of sulfur atoms arranged in a puckered or 'crown' conformation (Box 6.3). α-Octasulfur transitions to a β-polymorph, monoclinic sulfur, at 95.2 °C, leaving the puckered conformation unchanged. As the temperature increases, monoclinic sulfur melts to a yellow liquid, whose viscosity decreases until 155 °C and then increases dramatically, reaching a maximum at about 180 °C. The increase in viscosity is due to a reversible ROP that involves the homolytic cleavage of the eight-membered ring to form a biradical chain (Box 6.3). The biradical chain can add to itself, reforming the ring, or react with another ring, opening them to grow the chain by adding segments of eight sulfur atoms (Box 6.3). As the chain grows longer, it intertwines with each other, increasing the viscosity. A chain composed of 200 000 sulfur atoms can form at a temperature of maximum viscosity.[3] Quenching of the melt by pouring it into ice water yields a rubbery elastomer, characterised by a glass transition temperature of −30 °C. At room temperature the elastomer can eventually depolymerise and crystallise to the most thermodynamically stable allotrope, orthorhombic α-S_8. However, extraction of the rubbery elastomer with carbon disulfide yields polymeric sulfur characterised by a glass transition temperature of 75 °C.[67] The extraction process also removes the small equilibrium concentration of cyclo-octasulfur, which functions as a plasticiser and a crystallisation nucleus in the elastomer. The removal of the cyclo-octasulfur delays the depolymerisation onset and stabilises the elastomer. Alternatively, the elastomer can be stabilised by inverse vulcanisation. This process crosslinks the polymeric sulfur by reaction with a small organic molecule, typically a diene such as limonene and myrcene, creating a stable, functional polymeric material. The high sulfur content of polymeric sulfur imparts unique properties and creates opportunities for applications in lithium-ion batteries, IR-transparent lenses, and mercury capture.[68]

6.3 Polymers Containing Transition Metals

In this section, we will discuss examples of organometallic and coordination polymers containing transition metals (these include the inner transition metals). We distinguish between coordination and organometallic polymers in Chapter 1. In a coordination polymer, an organic ligand donates a lone pair of electrons to a metal atom or ion in the polymer scaffold. In contrast, in an organometallic polymer, a carbon atom covalently bonds to a metal atom in the polymer scaffold. A subclass of coordination polymers is the MOFs, which are composed of metal ions coordinated to organic ligands, forming porous one-, two- or three-dimensional extended, periodic structures. Notably, coordination and organometallic polymers may incorporate the main group metals; however, this sub-section will discuss only those containing transition metals.

The presence of transition metals within the polymer scaffold introduces bespoke designs or functionalities uncommon in organic and inorganic polymers. The transition metal can introduce variable oxidation states and magnetism, and enable interaction with electromagnetic radiation, leading to potential applications in catalysis, electronics, information storage, and photonics. Indeed, organometallic and coordination polymers are notable examples of functional polymers with many potential applications. Despite their promise, these polymers have delivered very little in real-world applications compared with organic and inorganic polymers. At the forefront of coordination polymers are MOFs, which, due to their desirable properties, are generating a lot of interest for potential applications in catalysis, electronics, photonics, drug delivery, and gas purification, separation, and storage. On the other hand, organometallic polymers such as polymetallocenes (polymers derived from metallocene monomers) are extensively investigated for potential applications in catalysis, coatings, electronics, and photonics. Whereas some MOFs are en route to commercialisation, polymetallocenes remain a laboratory curiosity with little or no commercial relevance at present. This section will discuss MOFs and polymetallocenes as prime examples of coordination and organometallic polymers, respectively.

6.3.1 Metal–Organic Frameworks

IUPAC defines metal–organic frameworks (MOFs) as coordination networks with organic ligands containing potential voids.[69] MOFs first garnered attention within the materials science community due to their ultrahigh porosity and surface area. A key advantage of MOFs over zeolites, a class of crystalline aluminosilicate materials, is the opportunity to design their porosity for specific applications. For example, the pore size depends on the length of the organic linker, so changing the length of the linker modifies the pore size. Also, selecting a particular type of functional group on the linker allows rational functionalisation of the pores. For a detailed discussion of the general synthesis, properties, and potential applications of MOFs, interested students should consult reviews and textbooks on this matter (see the Further Reading section).

The huge chemical space available allows the construction of diverse MOFs with over 30 000, including HKUST-1 (Figure 6.13), reported to date. Several companies are pursuing

(a)   (b)

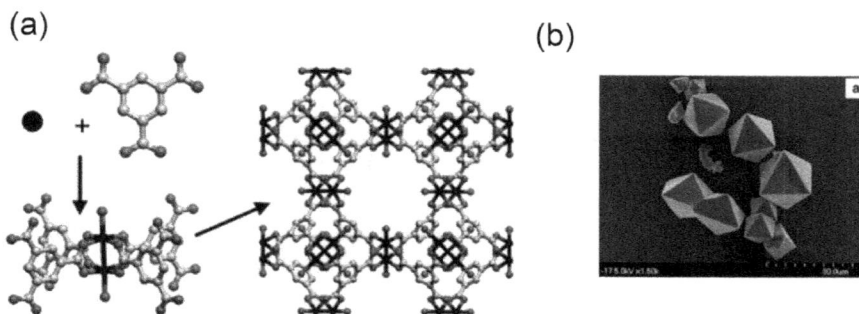

Figure 6.13 (a) The building blocks of HKUST-1. Two Cu^{2+} ions and four benzene-1,3,5 tricarboxylate ligands bond to give the MOF. Cu^{2+} is black, O is dark grey, and C is light grey (adapted with permission from ref. 76). (b) FE-SEM image of the as-synthesised HKUST-1 MOF. Reproduced from ref. 77 with permission from Elsevier, Copyright 2012.

proof-of-concept investigations to bring these materials to the market. For example, MOF Technologies, a spin-out company from Queen's University Belfast in the United Kingdom, has developed a carbon capture technology, Nuada, which combines vacuum swing adsorption with MOF-based filters to capture and remove carbon(IV) oxide at the emission source. HKUST-1, a copper-based coordination polymer $[Cu_3(TMA)_2(H_2O)_3]_n$ (TMA is trimesic acid [benzene-1,3,5-tricarboxylic acid]), is currently commercially available and sold by Sigma-Aldrich as Basolite™ C300. In HKUST-1, benzene-1,3,5-tricarboxylate bridges copper(II) paddlewheel dimers, forming face-centred-cubic crystals with a large square-shaped pore of 1 nm size and an accessible porosity of about 40% of the solid (Figure 6.13).[70] The paddlewheel is the secondary building unit of HKUST-1 and is built from four benzene-1,3,5-tricarboxylate molecules linking a dicopper(II) cluster (Figure 6.13). Each metal centre coordinates a water molecule in the hydrated state with the pore capable of containing up to ten additional water molecules per formula unit.[70] The water molecules can be replaced with a functional molecule such as pyridine to functionalise the pores. Also, HKUST-1 analogs can be obtained using other metals such as Mo^{2+},[71] Zn^{2+},[72] Ru^{2+},[73] Fe^{2+},[74] or Fe^{3+}.[75] Using a metal with a +2 oxidation state creates a neutral MOF, whereas a +3 oxidation state gives a positively charged framework, requiring a counteranion to achieve neutrality.

6.3.2 Polymetallocenes

Polymetallocenes are derived from metallocene (organometallic compounds in which a metal is bonded to a cyclopentadienyl group) monomers. After the serendipitous discovery of ferrocene (Box 6.2) – the first metallocene to be characterised – the synthesis and characterisation of ferrocene derivatives and other metallocenes and their derivatives motivated interest in metallocene-containing polymers.[8] Derivatives such as ferrocenylmethyl acrylate,[78] titanocene dichloride,[79] and ferrocenophanes[80] inspire the design of polymers containing metallocene moieties. The ROP of silicon-bridged [1]ferrocenophane yields high molecular weight poly(ferrocenylsilane) (Figure 6.14).[80] Currently, none of these polymers have commercial applications. Still, they exhibit exciting properties, including tunable redox activity and high refractive index, potentially valuable for diverse applications ranging from redox-responsive drug delivery to antireflection coatings (Box 6.5).

R = C$_6$H$_5$

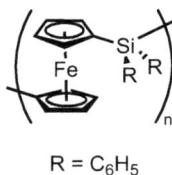

Figure 6.14   Structure of poly(ferrocenylsilane).

Box 6.5   Review questions.

1. Classify the polymers given below as inorganic, organometallic, and coordination polymers.

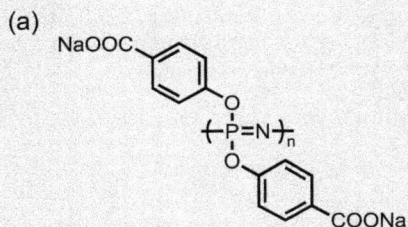

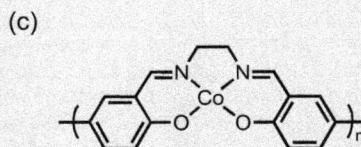

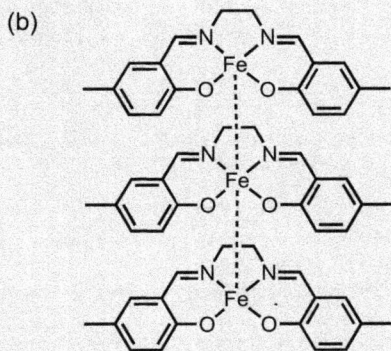

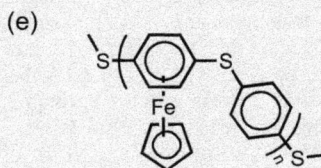

(a)

(c)

(b)

(d)

(e)

Further Reading

1. A. S. Abd-El-Aziz, C. Agatemor and W.-Y. Wong, *Macromolecules Incorporating Transition Metals: Tackling Global Challenges*, Royal Society of Chemistry Publishing, Cambridge, 2018.
2. J. E. Mark, H. R. Allcock and R. West, *Inorganic Polymers*, Oxford University Press Inc, New York, 2005.

3. R. D. Archer, *Inorganic and Organometallic Polymers*, Wiley-VCH, New York, 2001.
4. N. P. S. Chauhan and N. S. Chundawat, *Inorganic, and Organometallic Polymers*, Walter de Gruyter GmbH, Berlin, 2019.
5. G. L. Miessler, P. J. Fischer and D. A. Tarr, *Inorganic Chemistry*, Pearson Publishing, London, 5th edn, 2021.
6. C. Housecroft and A. G. Sharpe, *Inorganic Chemistry*, Pearson Publishing, London, 5th edn, 2018.
7. S. Kaskel, *The Chemistry of Metal–organic Frameworks: Synthesis, Characterization, and Applications*, Wiley-VCH, New York, 2016, vol. 1.

References

1. C. J. Giunta, V. V. Mianz and G. S. Girolami, *150 Years of the Periodic Table: A Commemorative Symposium (Perspectives on the History of Chemistry)*, Springer Nature, Switzerland, 1st edn, 2021.
2. C. E. Housecroft and A. G. Sharpe, *Inorganic chemistry*, Pearson Publishing, London, 5th edn, 2018.
3. G. Miessler, P. Fischer and D. Tarr, *Inorganic Chemistry*, Pearson Publishing, London, 5th edn, 2013.
4. S. A. Miller, J. A. Tebboth and J. F. Tremaine, *J. Chem. Soc.*, 1952, 632–635.
5. T. Kealy and P. Pauson, *Nature*, 1951, **168**, 1039–1040.
6. G. Wilkinson, P. Pauson, J. Birmingham and F. Cotton, *J. Am. Chem. Soc.*, 1953, **75**, 1011–1012.
7. G. Wilkinson, *J. Am. Chem. Soc.*, 1952, **74**, 6146–6147.
8. A. S. Abd-El-Aziz, C. Agatemor and N. Etkin, *Macromol. Rapid Commun.*, 2014, **35**, 513–559.
9. Y. Yan, J. Zhang, Y. Qiao, M. Ganewatta and C. Tang, *Macromolecules*, 2013, **46**, 8816–8823.
10. A. Gross, N. Hüsken, J. Schur, Ł. Raszeja, I. Ott and N. Metzler-Nolte, *Bioconjugate Chem.*, 2012, **23**, 1764–1774.
11. T. Matsumoto, Y. Onishi, K. Tanaka, H. Fueno, K. Tanaka and Y. Chujo, *Chem. Commun.*, 2014, **50**, 15740–15743.
12. Y. Qin, V. Sukul, D. Pagakos, C. Cui and F. Jäkle, *Macromolecules*, 2005, **38**, 8987–8990.
13. N. Brasseur, R. Ouellet, C. La Madeleine and J. Van Lier, *Br. J. Cancer*, 1999, **80**, 1533–1541.
14. J. Koe, *Polym. Int.*, 2009, **58**, 255–260.
15. T. Hayashi, Y. Uchimaru, N. P. Reddy and M. Tanaka, *Chem. Lett.*, 1992, **21**, 647–650.
16. W.-M. Zhou and I. Tomita, *J. Inorg. Organomet. Polym. Mater.*, 2009, **19**, 113–117.
17. M. Richards, B. Dahiyat, D. Arm, P. Brown and K. Leong, *J. Biomed. Mater. Res.*, 1991, **25**, 1151–1167.
18. M. Grosche, E. Herdtweck, F. Peters and M. Wagner, *Organometallics*, 1999, **18**, 4669–4672.
19. H. Imoto and K. Naka, *Polymer*, 2022, 124464.
20. H. Hu, L. Wang, L. Wang, L. Li and S. Feng, *Polym. Chem.*, 2020, **11**, 7721–7728.
21. Y. Kobayashi, A. Harada and H. Yamaguchi, *Chem. Commun.*, 2020, **56**, 13619–13622.
22. Q. Li, K. L. Ng, X. Pan and J. Zhu, *Polym. Chem.*, 2019, **10**, 4279–4286.
23. L. Kloo, Catenated Compounds – Group 17 – Polyhalides, in *Comprehensive Inorganic Chemistry II: From Elements to Applications*, Elsevier, 2nd edn, 2013, pp. 233–249.
24. M. C. Aragoni, M. Arca, S. J. Coles, F. A. Devillanova, M. B. Hursthouse, S. L. Coles, F. Isaia, V. Lippolis and A. Mancini, *CrystEngComm*, 2011, **13**, 6319–6322.
25. K. M. Fromm, *Coord. Chem. Rev.*, 2008, **252**, 856–885.
26. P. J. Fazen, J. S. Beck, A. T. Lynch, E. E. Remsen and L. G. Sneddon, *Chem. Mater.*, 1990, **2**, 96–97.
27. K. Su, E. E. Remsen, H. M. Thompson and L. G. Sneddon, *Macromolecules*, 1991, **24**, 3760–3766.
28. T. Wideman and L. G. Sneddon, *Chem. Mater.*, 1996, **8**, 3–5.
29. M. Mutailipu, K. R. Poeppelmeier and S. Pan, *Chem. Rev.*, 2020, **121**, 1130–1202.
30. Ü. Sızır, Ö. Yurdakul, D. A. Köse and F. Akkurt, *Molecules*, 2019, **24**, 2790.
31. M. A. Beckett, B. I. Meena, T. A. Rixon, S. J. Coles and P. N. Horton, *Molecules*, 2019, **25**, 53.
32. M. A. Beckett, *Coord. Chem. Rev.*, 2016, **323**, 2–14.
33. H. L. van de Wouw, E. C. Awuyah, J. I. Baris and R. S. Klausen, *Macromolecules*, 2018, **51**, 6359–6368.
34. D. A. Resendiz-Lara, G. R. Whittell, E. M. Leitao and I. Manners, *Macromolecules*, 2019, **52**, 7052–7064.
35. A. Schäfer, T. Jurca, J. Turner, J. R. Vance, K. Lee, V. A. Du, M. F. Haddow, G. R. Whittell and I. Manners, *Angew. Chem.*, 2015, **127**, 4918–4923.
36. C. Marquardt, T. Jurca, K. Schwan, A. Stauber, A. V. Virovets, G. R. Whittell, I. Manners and M. Scheer, *Angew. Chem., Int. Ed.*, 2015, **54**, 13782–13786.

37. H. Dorn, J. M. Rodezno, B. Brunnhöfer, E. Rivard, J. A. Massey and I. Manners, *Macromolecules*, 2003, **36**, 291–297.
38. D. A. Resendiz-Lara, V. T. Annibale, A. W. Knights, S. S. Chitnis and I. Manners, *Macromolecules*, 2020, **54**, 71–82.
39. A. Samokhvalov, *Coord. Chem. Rev.*, 2018, **374**, 236–253.
40. N. Y. Gugin, A. Virovets, E. Peresypkina, E. I. Davydova and A. Y. Timoshkin, *CrystEngComm*, 2020, **22**, 4531–4543.
41. T. Rabe, E. S. Grape, H. Rohr, H. Reinsch, S. Wöhlbrandt, A. Lieb, A. K. Inge and N. Stock, *Inorg. Chem.*, 2021, **60**, 8861–8869.
42. T. C. Schäfer, J. Becker, D. Heuler, M. T. Seuffert, A. E. Sedykh and K. Müller-Buschbaum, *Aust. J. Chem.*, 2022, **75**, 676–683.
43. R. Munirathinam, R. Ricciardi, R. J. Egberink, J. Huskens, M. Holtkamp, H. Wormeester, U. Karst and W. Verboom, *Beilstein J. Org. Chem.*, 2013, **9**, 1698–1704.
44. C. He, J. Dong, C. Xu and X. Pan, *ACS Polym. Au*, 2023, **3**, 5–27.
45. Y. Qin, G. Cheng, A. Sundararaman and F. Jäkle, *J. Am. Chem. Soc.*, 2002, **124**, 12672–12673.
46. M. Alvaro, C. Baleizao, E. Carbonell, M. El Ghoul, H. García and B. Gigante, *Tetrahedron*, 2005, **61**, 12131–12139.
47. G. Odian, *Principles of polymerization*, John Wiley & Sons, 2004.
48. J. Ibemesi, N. Gvozdic, M. Kuemin, Y. Tarshiani and D. Meier, *MRS Online Proc. Libr.*, 1989, **171**, 105.
49. H. Kakiuchida, M. Takahashi, Y. Tokuda, H. Masai and T. Yoko, *J. Phys. Chem. B*, 2007, **111**, 982–988.
50. Y. Zhang, L. Yuan, G. Liang and A. Gu, *J. Mater. Chem. A*, 2018, **6**, 23425–23434.
51. M. Hara, Y. Iijima, S. Nagano and T. Seki, *Sci. Rep.*, 2021, **11**, 1–8.
52. H. Gilman and W. H. Atwell, *J. Organomet. Chem.*, 1965, **4**, 176–178.
53. H. Gilman, W. H. Atwell and G. L. Schwebke, *J. Organomet. Chem.*, 1964, **2**, 369–371.
54. R. D. Miller and J. Michl, *Chem. Rev.*, 1989, **89**, 1359–1410.
55. P. A. Bianconi, F. C. Schilling and T. W. Weidman, *Macromolecules*, 1989, **22**, 1697–1704.
56. C. Aitken, J. F. Harrod and E. Samuel, *Can. J. Chem.*, 1986, **64**, 1677–1679.
57. G. Barroso, M. Döring, A. Horcher, A. Kienzle and G. Motz, *Adv. Mater. Interfaces*, 2020, **7**, 1901952.
58. D. Amouzou, L. Fourdrinier, F. Maseri and R. Sporken, *Appl. Surf. Sci.*, 2014, **320**, 519–523.
59. L. Picard, P. Phalip, E. Fleury and F. Ganachaud, *Prog. Org. Coat.*, 2015, **80**, 120–141.
60. R. H. Neilson and P. Wisian-Neilson, *Chem. Rev.*, 1988, **88**, 541–562.
61. P. Wisian-Neilson and R. H. Neilson, *J. Am. Chem. Soc.*, 1980, **102**, 2848–2849.
62. C. H. Honeyman, I. Manners, C. T. Morrissey and H. R. Allcock, *J. Am. Chem. Soc.*, 1995, **117**, 7035–7036.
63. H. R. Allcock, C. A. Crane, C. T. Morrissey, J. M. Nelson, S. D. Reeves, C. H. Honeyman and I. Manners, *Macromolecules*, 1996, **29**, 7740–7747.
64. S. Rothemund and I. Teasdale, *Chem. Soc. Rev.*, 2016, **45**, 5200–5215.
65. N. Çini and V. Ball, *Adv. Colloid Interface Sci.*, 2014, **209**, 84–97.
66. T. V. Kulakovskaya, V. M. Vagabov and I. S. Kulaev, *Process Biochem.*, 2012, **47**, 1–10.
67. N. H. Ray, *Inorganic Polymers*, Academic Press, 1978.
68. D. Parker, H. Jones, S. Petcher, L. Cervini, J. Griffin, R. Akhtar and T. Hasell, *J. Mater. Chem. A*, 2017, **5**, 11682–11692.
69. S. R. Batten, N. R. Champness, X.-M. Chen, J. Garcia-Martinez, S. Kitagawa, L. Öhrström, M. O'Keeffe, M. P. Suh and J. Reedijk, *Pure Appl. Chem.*, 2013, **85**, 1715–1724.
70. S. S.-Y. Chui, S. M.-F. Lo, J. P. Charmant, A. G. Orpen and I. D. Williams, *Science*, 1999, **283**, 1148–1150.
71. M. Kramer, U. Schwarz and S. Kaskel, *J. Mater. Chem.*, 2006, **16**, 2245–2248.
72. J. I. Feldblyum, M. Liu, D. W. Gidley and A. J. Matzger, *J. Am. Chem. Soc.*, 2011, **133**, 18257–18263.
73. W. Zhang, K. Freitag, S. Wannapaiboon, C. Schneider, K. Epp, G. Kieslich and R. A. Fischer, *Inorg. Chem.*, 2016, **55**, 12492–12495.
74. Y. Yue, H. Arman, Z. J. Tonzetich and B. Chen, *Z. Anorg. Allg. Chem.*, 2019, **645**, 797–800.
75. S. Sotnik, S. Kolotilov, M. Kiskin, Z. V. Dobrokhotova, K. Gavrilenko, V. Novotortsev, I. Eremenko, V. Imshennik, Y. V. Maksimov and V. Pavlishchuk, *Russ. Chem. Bull.*, 2014, **63**, 862–869.
76. S. Bordiga, L. Regli, F. Bonino, E. Groppo, C. Lamberti, B. Xiao, P. Wheatley, R. Morris and A. Zecchina, *Phys. Chem. Chem. Phys.*, 2007, **9**, 2676–2685.
77. K.-S. Lin, A. K. Adhikari, C.-N. Ku, C.-L. Chiang and H. Kuo, *Int. J. Hydrogen Energy*, 2012, **37**, 13865–13871.
78. C. U. Pittman Jr, J. Lai, D. Vanderpool, M. Good and R. Prado, *Macromolecules*, 1970, **3**, 746–754.
79. C. E. Carraher, *J. Inorg. Organomet. Polym. Mater.*, 2005, **15**, 121–145.
80. D. A. Foucher, B. Z. Tang and I. Manners, *J. Am. Chem. Soc.*, 1992, **114**, 6246–6248.

7 Applications of Inorganic and Organometallic Polymers

7.1 Introduction

In the preceding chapters, we learned the properties of inorganic and organometallic polymers. These properties enable many commercial and potential applications, which will be discussed in this chapter. Because these applications depend on the properties of these polymers, we recommend that the student revisit Chapter 5 to grasp the fundamental properties of inorganic and organometallic polymers before reading this chapter. It is also crucial that the student appreciates that polymers need to be processed into plastics, elastomers, fibers, or coatings (see Chapter 1) before use. Therefore, these processed forms should not be confused with the applications; instead, we will discuss real or potential applications, such as drug delivery and sunlight harvesting. We will discuss potential applications because, as discussed in the preceding chapters, most inorganic and organometallic polymers are yet to enter the market. To facilitate a discussion that is understandable to scientists and non-scientists, we will structure our discussion in this chapter around applications in harsh environments, healthcare, the chemical industry, and sustainable energy (Box 7.1).

Box 7.1 Learning outcomes.

By the end of this chapter, the student should be able to:
1. Identify inorganic polymers in commercial applications.
2. Propose applications for inorganic and organometallic polymers based on their properties.
3. Discuss properties required of polymers used in tissue engineering.
4. Discuss properties required of polymers to function through their lifetime.

7.2 Applications in Harsh Environments

During their expected lifetime of use, materials, whether metals, ceramics, or polymers, need to preserve their properties. This requirement means that a stiff material should remain stiff, while a flexible material should retain its flexibility throughout its lifetime. This requirement is critical to materials used under harsh conditions. These conditions include the abnormal temperatures under which aeronautical vehicles operate, the hydrolytic environments under which marine and submarine vessels operate, and intense radiations in nuclear reactors. Designing materials for these harsh environments is an extraordinary challenge and often requires multifaceted structural optimisations, such as alloying, doping, composite fabrication, or copolymerisation. Currently, metallic alloys and ceramics, due to their thermal stability, durability, and mechanical strength, constitute most of the materials used in harsh environments.

In designing materials for use in these environments, some properties are of interest. For instance, thermal stability is critical to the robust performance of materials designed for aeronautical vehicles. Water repellency and resistance to hydrolytic degradation and biofilm formation are properties needed in materials intended for construction of marine and submarine vessels (Table 7.1). Many thermo-oxidatively stable organic polymers, such as polyimides and polyamides, exhibit properties suited for use in harsh environments. Nonetheless, inorganic polymers, such as polysiloxanes and polyphosphazenes (see Chapter 6), are promising alternatives to fabricate materials intended for use in harsh environments. As an example, polyfluoroalkoxyphosphazenes (Figure 7.1) are elastic over a wide temperature range of −95 to 270 °C and resist chemical attack, implying a potential use as a cryogenic fuel handling hose. These properties are useful for materials designed for aircraft that operate over a broad temperature range (−55 °C to 50 °C at the ground level and down to −80 °C at high altitude). Also, several polyorganophosphazene copolymers have flame retardancy and water repellency properties, with poly(aryloxy)phosphazenes being potentially useful as thermal insulation lining for the interior of submarine hulls.

For aerospace applications, the need for chemically-resistant surfaces is becoming relevant with the increasing use of chemicals for disinfecting aircraft interior surfaces. These disinfectants must meet the stringent flame, smoke, toxicity, and heat-release standards mandated for commercial aircrafts. Because of their unique structure,

Table 7.1 Properties required of materials intended for use in harsh environments.[a]

	Aerospace	Marine
Adverse forces	Stress Abnormal temperatures Oxidation	Abnormal temperatures Hydrolysis Stress Fouling
Properties required of materials	Thermal stability Oxidative stability Ablation resistance Low density High Young's modulus	Antifouling Water repellency Thermal stability Hydrolytic stability

[a]This is from the materials property perspective. Performance, manufacturability, and environmental safety perspectives may require different properties.

(a)   (b)

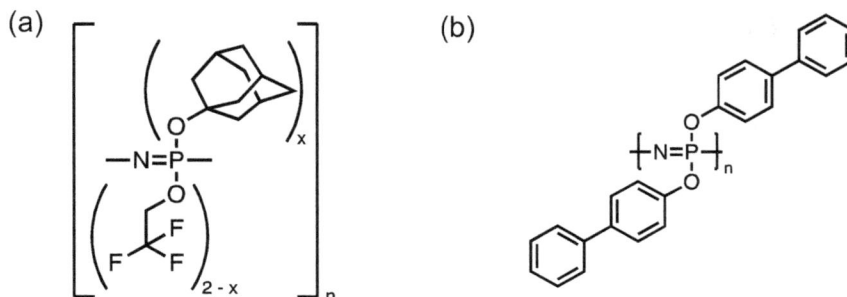

Figure 7.1 Structural depiction of (a) polyfluoroalkoxyphosphazenes and (b) poly(aryloxy)-phosphazenes.

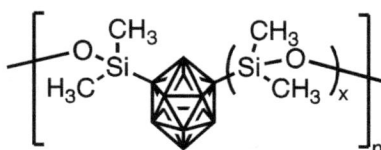

Figure 7.2 Structural depiction of the polysiloxane–carborane copolymer (Ucarsil®).

polysiloxane-based adhesives exhibit excellent heat resistance that may satisfy the requirements mandated for commercial aircrafts. A caveat is that polysiloxane adhesives degrade in a polar medium at a temperature above 200 °C or in an oxygen-rich environment. A polysiloxane–carborane copolymer (Ucarsil®) (Figure 7.2) developed by Olin Matheson and Princeton Laboratories in the mid-1960s features superior thermal stability (300 °C in the air), supporting ultrahigh temperature applications as an elastomer. Compared to thermally stable organic polymers such as polytetrafluoroethylene or inorganic polymers such as polysiloxanes, Ucarsil® combines a unique repertoire of properties, including excellent processability and fire and chemical resistance, which is well-suited for fabrication of materials that meet the stringent requirement in the aerospace industry. These polymers also retain their mechanical properties down to −40 °C.

7.3 Applications in Healthcare

Inorganic and organometallic polymers remain attractive biomaterials with potential applications as drugs, drug delivery vectors, biosensors, bioimaging probes, and tissue engineering scaffolds. The application of inorganic and organometallic polymers in biomedicine is due to the following:

(a) the functional properties of constituent inorganic elements or transition metals, and
(b) polymer properties such as molecular weight and architecture.

While polymer properties are general to all polymers, the functional properties, including redox activity, photoactivity, and coordination chemistry, are unique to inorganic and organometallic polymers. These functional properties sometimes regulate biological processes, endowing these polymers with therapeutic or antipathogenic properties. In other cases, the functional properties respond to changes in biological conditions, enabling the polymer to act as a biosensor or bioimaging probe. Also, the biological environment tunes the functional properties to alter the properties of the polymer such as self-assembly or solubility, allowing applications in controlled drug delivery. Although most organometallic polymers are bioactive, perturbing biological processes, several inorganic polymers are biocompatible, allowing their use in drug delivery and tissue engineering. This section will discuss several inorganic and organometallic polymers with demonstrated potential in treating diseases, mitigating drug-resistant infections, targeted drug delivery of therapeutic agents, biosensing and bioimaging, and regenerative medicine.

7.3.1 Application as Drugs

A search of the DrugBank database at www.drugbank.com shows several approved organometallic compounds indicated for treating or detecting diseases. An organometallic polymer is yet to be indicated as a drug. However, the potential exists for applying the polymeric analogs of these approved drugs, such as cisplatin, in treating diseases. Polymers are a platform technology to introduce multivalent interactions between multiple ligands and their corresponding endogenous receptors, target drugs to the desired tissue/organ, mitigate side effects, and combine the therapeutic benefits of multiple drugs. To illustrate, cisplatin, nedaplatin, carboplatin, and oxaliplatin are platinum-based chemotherapy drugs with proven efficacy in treating several cancers. Despite their proven success, these platinum-based anticancer drugs have several adverse side effects, notably nephrotoxicity, myelosuppression, ototoxicity, and neurotoxicity. A strategy to mitigate the systemic toxicity of these platinum-based drugs is selective targeting to the tumor. In the late 1990s, Duncan and coworkers designed a polymeric analog of cisplatin (Figure 7.3). They grafted cisplatin to a biocompatible and water-soluble polymer, namely, the *N*-(2-hydroxy-propyl)methacrylamide (HPMA) copolymer, to reduce the systemic toxicity and localise cisplatin in the tumor.[1] Compared with molecular cisplatin, the polymer analog induces a 5–15-fold decrease in toxicity and results in a 60-fold increase in platinum concentration in the tumor of the mouse model of melanoma.[1] Access Pharmaceuticals acquired the technology, developing the polymer analogs of carboplatin (AP5280) and oxaliplatin (AP5346) which were investigated in phase I/II clinical trials in patients with recurrent ovarian cancer.[2] Pharmacodynamic studies in preclinical models of B16 and other tumors show that AP5346, later renamed ProLindac™, delivers 16 times more platinum to tumors and results in 13 times more platinum–tumor DNA complexes compared with oxaliplatin at equitoxic doses. The mechanism of action of AP5346 involves the localisation of the polymer–drug conjugate in the tumor by the enhanced permeability and retention effect, followed by the low pH-triggered release of the drug, diaminocyclohexane platinum, to the tumor. Once in the tumor, platinum binds preferentially to the guanine and cytosine moieties of DNAs, resulting in DNA crosslinking and complexation and ultimately inhibiting DNA biosynthesis and biological function.

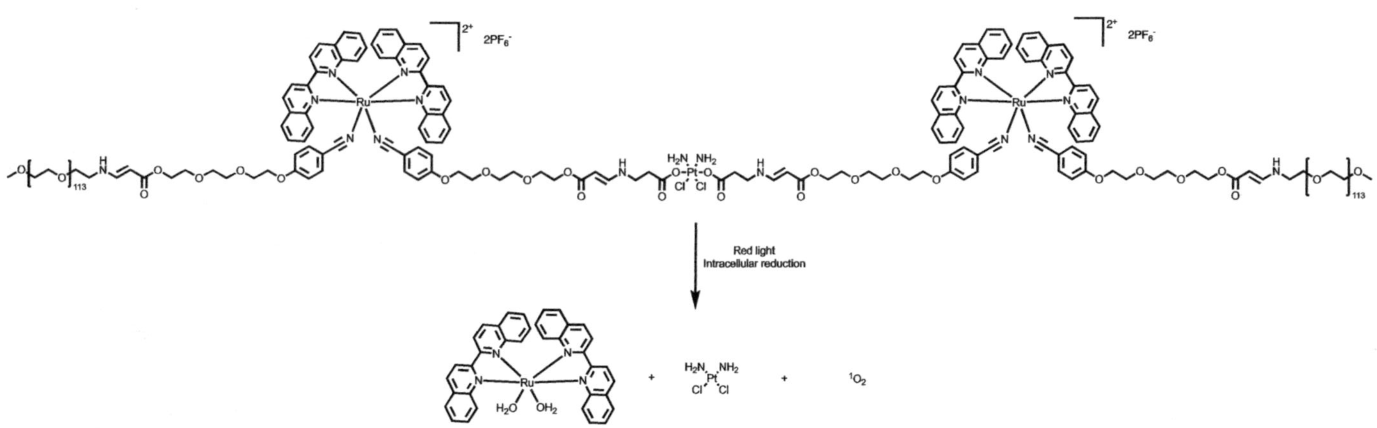

Figure 7.3   Structure of the polymeric drug incorporating cisplatin and [Ru(biq)$_2$(H$_2$O)$_2$][(PF$_6$)$_2$ (biq = 2,2'-biquinoline) in its backbone. Red light irradiation and intracellular reduction release the anticancer cisplatin and [Ru(biq)$_2$(H$_2$O)$_2$](PF$_6$)$_2$ and singlet oxygen that generates ROS.

ProLindac™ is better described as an organic polymer–organometallic drug conjugate.[3] Polymeric drugs are another subclass of inorganic and organometallic polymer-based drugs, whereby the polymer backbone comprises a therapeutic organometallic moiety. The therapeutic properties of the organometallic moiety stem from the following:

(a) its coordination chemistry that might lead to undesirable interactions with endogenous molecules such as DNAs,
(b) its redox chemistry that could culminate in the generation of reactive oxygen species (ROS), leading to oxidative stress, or
(c) photoactivity that could also lead to the generation of ROS.

The polymer framework provides an excellent platform to integrate or synergise the three mechanisms to develop a potent multimodality, also called combination therapy, to treat monotherapy-resistant diseases. A polymeric drug containing two organometallic anticancer agents, cisplatin and $[Ru(biq)_2(H_2O)_2](PF_6)_2$ (biq = 2,2'-biquinoline), in its backbone, typifies a multimodal therapy that can attenuate the growth of cisplatin-resistant tumors in a mouse model (Figure 7.3).[4] The polymer amphiphilicity and the therapeutic properties of the organometallic drug enhance the efficacy. The polymer amphiphilicity leads to self-assembled nanoparticles that localise in the tumor *via* EPR effects. Red light irradiation and intracellular reduction degrade the polymer nanoparticles, generating ROS and releasing the anticancer drugs. The ROS, cisplatin, and $[Ru(biq)_2(H_2O)_2](PF_6)_2$ exhibit a distinct anticancer mechanism that synergises to inhibit the growth of cisplatin-resistant cancer cells.

The modalities discussed above exemplify chemotherapy using cytotoxic chemicals to inhibit cell division or induce DNA damage. Chemotherapy constitutes a part of systemic therapy that includes immunotherapy and can be combined with local therapy, such as radiotherapy or photodynamic therapy, to treat drug-resistant diseases. Coordination polymers, designed from photoactive or radioactive transition metals and possessing porous cores, are an excellent platform for combination therapy. Chemotherapy/immunotherapy can be encapsulated within the porous core, while the transition metal can be irradiated to kill cancer cells. Indeed, metal–organic frameworks (MOFs), a class of coordination polymers (see Chapters 1 and 6), can be designed to possess the porosity to encapsulate chemotherapy/immunotherapy and photoactive or radioactive transition metals to generate reactive oxygen species to eradicate cancer cells.[5–7] MOFs, such as hafnium 5,15-di(*p*-benzoato)porphyrin and hafnium 5,10,15,20-tetra(*p*-benzoato)porphyrin, are typical examples of coordination polymer platforms for combination therapy.[6] The hafnium cluster in the MOFs absorbs X-ray photons to generate ROS, enabling radiotherapy-based treatment.[6] Simultaneously, the porous core of the MOFs encapsulates a small molecule immune checkpoint inhibitor to re-sensitise cancer cells to T cell immunosurveillance.[6]

7.3.2 Application as Antipathogenic Agents

Inorganic and organometallic compounds have a long history of antipathogenic activity and are receiving renewed attention owing to the emergence of drug-resistant infections. Undoubtedly, inorganic and organometallic compounds are essential to the survival of all living organisms including pathogens, but they become toxic in certain forms and at

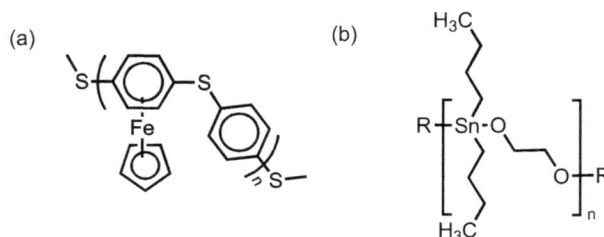

Figure 7.4 Structure of (a) an antibacterial organometallic polymer with activity against methicillin-resistant *Staphylococcus aureus* and (b) an antiviral organometallic polymer.

certain concentrations. The exact mechanism of the antipathogenic activity of these compounds is multifaceted, varies significantly between metal complexes, and depends on the types of the metal, ligand, and coordination sphere.[8] The synthesis of polymers incorporating antipathogenic inorganic and organometallic moieties provides precursors to fabricate antipathogenic and antifouling materials with the potential to mitigate the spread of drug-resistant infections.[9–11] Several metal-containing polymers have been shown to inhibit the growth of bacteria, fungi, and viruses (Figures 7.4 and 7.5). Also, these polymers are being proposed as potential candidates to be used alongside traditional antibiotics in combination therapy (see Section 7.3.1). The additive or synergistic effect of antipathogenic polymers may improve the treatment efficacy of antibiotics by rendering the pathogens more susceptible to antibiotics.[12] Like organometallic complexes, the exact mechanism of action of these polymers may be convoluted. Multiple mechanisms may cooperate to exert antipathogenic activity. These mechanisms include generating sublethal doses of ROS, reducing ATP levels, and disrupting membrane integrity.

7.3.3 Application as Controlled Drug Delivery Systems

The organometallic polymeric drug described above is, strictly speaking, a prodrug because the pharmaceutically active organometallic moiety is released upon polymer degradation. Indeed, these polymers are archetypal targeted drug delivery systems, capable of releasing the conjugated organometallic prodrug upon stimulation by an exogenous or endogenous trigger. However, the functional properties, such as redox activity, of organometallic polymers can be exploited for targeted delivery of drugs encapsulated within the core of organometallic polymer-based hydrogels and nanomaterials. To illustrate, the redox-active organometallic compound, ferrocene, changes from a hydrophobic to a hydrophilic molecule upon oxidation to the ferrocenium ion. This transition profoundly affects the gel–sol transition of hydrogels derived from ferrocene-containing polymers or self-assembly/disassembly properties of nanomaterials derived from ferrocene-containing polymers (Figure 7.6). An example is the gel–sol transition of the hydrogel derived from the complexation of cyclodextrin-modified poly(acrylic acid) and ferrocene-modified poly(acrylic acid) upon oxidation of the ferrocene to ferrocenium ion. Such gel–sol transition has far-reaching implications for controlled drug delivery, where a hydrogel-encapsulated drug can be released upon activating a redox trigger. The same applies to a ferrocene-containing polymeric

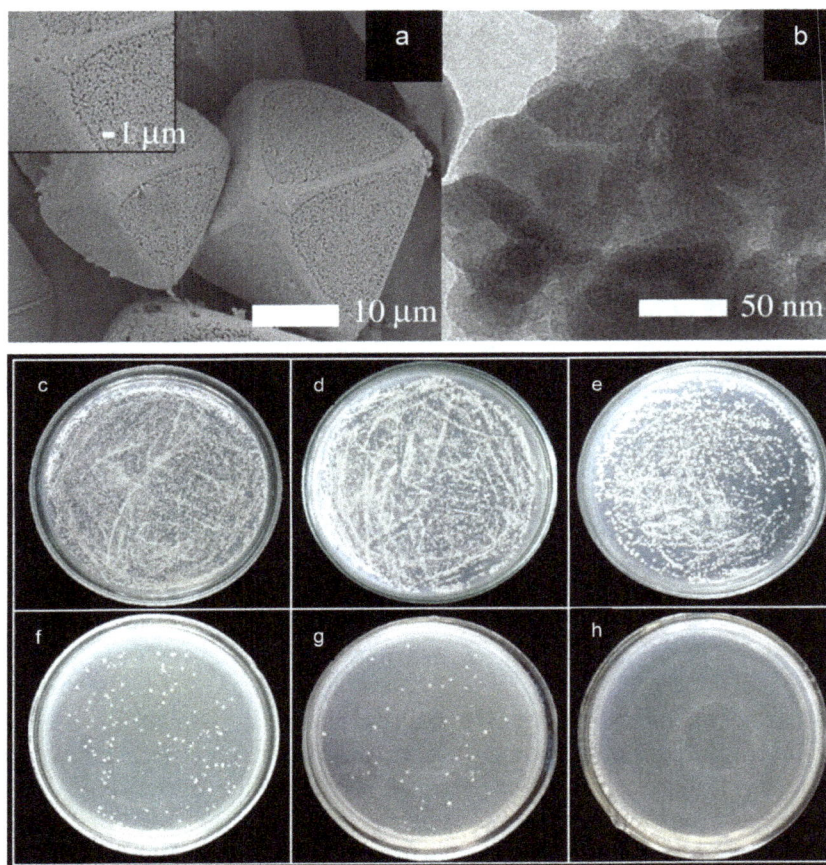

Figure 7.5 (a) Field-emission scanning electron microscopy and (b) transmission electron microscopy images of an antifungal copper-1,3,5-benzenetricarboxylate metal–organic framework (Cu-BTC MOF) with activity against *Candida albicans, Aspergillus niger, Aspergillus oryzae,* and *Fusarium oxysporum.*[13] (c–h) Antifungal properties of the Cu-BTC MOF at different concentrations against *C. albicans* after 60 minutes of ncubation: (c) control, (d) 100 ppm, (e) 200 ppm, (f) 300 ppm, (g) 400 ppm, and (h) 500 ppm. Reproduced from ref. 13, https://doi.org/10.1098/rsos.170654, under the terms of the CC BY 4.0 license, https://creativecommons.org/licenses/by/4.0/.

nanomaterial that can exhibit redox-triggered assembly and disassembly to encapsulate and release a drug. For example, redox-responsive micelles designed through the host–guest interaction between β-cyclodextrin grafted dextran and ferrocene-terminated poly(ε-caprolactone) enable the controlled release of meloxicam, a practically water-insoluble NSAID drug.[14]

Apart from organometallic polymers exemplified by ferrocene-containing polymers, inorganic polymers such as polyphosphazenes and polysiloxanes feature properties suitable for drug delivery. Polyphosphazenes, for instance, feature tunable biodegradable rates, bioavailability, biocompatibility, tailorable structural architecture, and nontoxic degradation by-products. The ability of polymers to degrade to nontoxic by-products is highly desirable in drug delivery. Polyphosphazenes containing ester linkages, for example, degrade to the corresponding alcohol, amine, ammonia, and phosphate,

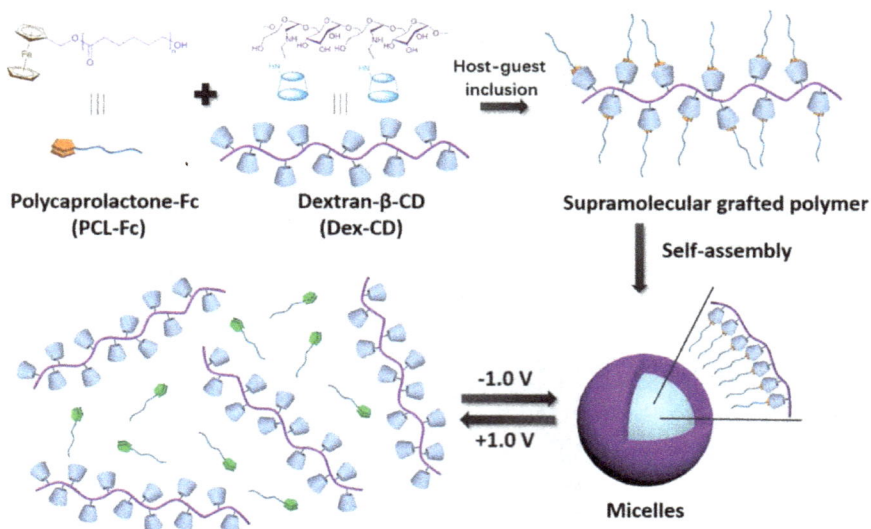

Figure 7.6   Schematic depiction of redox-triggered self-assembly and disassembly of micelles formed from redox active ferrocene-containing polycaprolactone and β-cyclodextrin-containing dextran. Reproduced from ref. 14 with permission from American Chemical Society, Copyright 2019.

products that are less likely to perturb cellular homeostasis at low concentrations. A caveat is that high concentrations of ammonia are cytotoxic. Therefore, the bio-compatibility of polyphosphazenes is concentration dependent. Nonetheless, polyphosphazenes remain potentially useful as biomaterials for formulating diverse nanomaterials such as micelles, vesicles, hydrogels, polyelectrolytes, and polymersomes for controlled drug delivery. These nanomaterials can be formulated by exploiting the tailorable architecture of polyphosphazenes. For example, grafting poly(ethylene glycol) to polyphosphazene yields an amphiphile that self-assembles into micelles[15] or poly-mersomes[16,17] capable of encapsulating therapeutics that can be released in a sustained manner. Polymers, molecules, and drugs can be simultaneously grafted onto the poly-phosphazene backbone to tailor the properties of the resulting polymer. Simultaneously grafting a hydrophilic poly(ethylene glycol) and hydrophobic L-isoleucine ethyl ester on a polyphosphazene–doxorubicin conjugate yields a biodegradable, thermosensitive hydrogel that locally releases the doxorubicin over an extended time.[18]

Polysiloxanes are another archetype of inorganic polymers used in drug delivery. They have low modulus, good thermal stability, non-adhesiveness, oxidative and chemical stability, and excellent biocompatibility. It is essential to understand that biocompatibility depends on bulk and interfacial properties. Bulk properties, such as modulus and thermal stability, dictate how the material interacts mechanically with the surrounding tissues. On the other hand, interfacial properties, such as adhesive-ness and bioinertness, are related to how the surface chemical composition of the material interacts with surrounding tissues. It is also crucial to understand that the duration and dynamics of use impact biocompatibility, with applications over a long period likely to fail, degrade or cause problems. Provided the interfacial properties are acceptable for biomedical applications, issues originating from the bulk properties of the polymer can be alleviated by design and engineering. For example, changing the molecular weight and degree of crosslinking can tune the mechanical properties. Polysiloxanes have acceptable bulk and interfacial properties, informing their ongoing drug-delivery exploration. Indeed, polydimethylsiloxanes have been formulated into nanoparticles to encapsulate and release doxorubicin.[19] Also, several hyperbranched polysiloxanes have been designed for high drug loading and pH-responsive release of the hydrophobic ibuprofen.[20,21] It is important to note that polysiloxanes biodegrade with the high molecular weight polymer degrading at a lower rate than low molecular weight analogs.[22,23]

7.3.4 Application as Biosensors and Bioimaging Probes

Biosensing and bioimaging are subdisciplines of bioanalytical chemistry concerned with developing modalities to investigate biological processes at the molecular level. In biosensing, a device provides qualitative or quantitative information on biological analytes or structures, while a bioimaging modality generates a visual representation of biological analytes or structures. Biological analytes or structures can interact with inorganic and organometallic polymers, modifying the functional or structural properties of the polymers to generate information to detect the levels or identity of the analytes or structure. Among these are optical and electrochemical properties, which are present in most inorganic and organometallic polymers. For instance, polysiloxane aggregates exhibit intrinsic fluorescence, an optical property potentially valuable for bioimaging and biosensing (Figure 7.1).[20,21,24,25] This intrinsic fluorescence is explained by aggregation-induced emission (AIE). The phenomenon of AIE occurs when materials show enhanced photoluminescence in the aggregated state compared with the solution state. The intrinsic fluorescence of polysiloxanes is improved by changing the amphiphilicity of the polymer by grafting hydrophilic or hydrophobic molecules as pendant to the silicone–oxygen backbone. The change in amphiphilicity reorganises the microstructure of the self-assembled aggregates, tuning the electronic properties for enhanced fluorescence as observed in a hyperbranched polysiloxane-*graft*-cyclodextrin. Theoretical calculations and transmission electron microscopy data showed that, in these graft polysiloxanes, the synergy between hydrogen bonds and hydrophobic interactions induces the formation of large aggregates and promotes through-space electron delocalisation systems.[21] The biocompatibility and the intrinsic AIE properties make this polysiloxane-*graft*-cyclodextrin a valuable bioimaging material. Indeed, this polysiloxane-*graft*-cyclodextrin enables visualisation of the cytoplasm of mouse osteoblast cells (Figure 7.7).[21]

Interactions between biological analytes and structures can change the electronic states, leading to luminescence, which can be detected for biosensing. Indeed, the AIE properties of a hyperbranched polysiloxane-*graft*-oleic acid are sensitive to the presence of various biologically relevant metals, specifically Fe^{3+}, Ba^{2+}, Ca^{2+}, Cd^{2+}, Co^{2+}, Cu^{2+}, Hg^{2+}, Na^+, Zn^{2+}, and Al^{3+}.[20] Particularly, Fe^{3+} significantly quenches the fluorescence intensity compared with Ba^{2+}, Ca^{2+}, Cd^{2+}, Co^{2+}, Cu^{2+}, and Hg^{2+}, whereas Na^+, Zn^{2+}, and Al^{3+} enhance the intensity.[20] The enhanced quenching of the AIE by Fe^{3+} suggests this polysiloxane may selectively detect Fe^{3+}. Overall, the non-classical AIE exhibited by some inorganic polymers, such as polysiloxanes, is potentially useful in biosensing

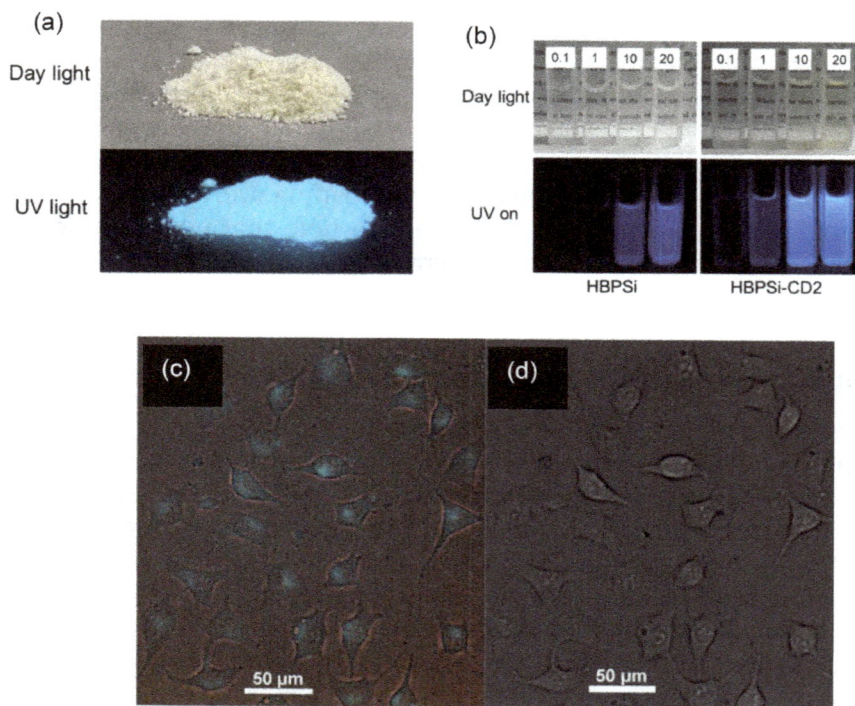

Figure 7.7   (a and b) Digital picture of β-cyclodextrin-containing hyperbranched polysiloxane showing aggregation-induced fluorescence under 365 nm UV light in the solid state and in aqueous solution with different concentrations in the range of 0.1–20 mg mL^{-1}. (c and d) Laser scanning confocal microscopy images of mouse osteoblasts after incubation in cell culture medium spiked with 2 mg mL^{-1} of β-cyclodextrin-containing hyperbranched polysiloxane. (c) Under excitation at 320–380 nm (d) and bright field. Reproduced from ref. 21 with permission from American Chemical Society, Copyright 2019.

and bioimaging applications. These applications are also doable with transition metal-containing polymers due to the classical photoluminescence behavior of transition metals (see Chapter 5). For example, troponin, a cardiac biomarker, selectively quenches and blue-shifts the photoluminescence of a manganese-containing MOF (Figure 7.1).[26] Applying the Stern–Völmer equation shows that the relationship between the concentration of troponin and photoluminescence quenching is linear over a wide concentration range. This linearity implies the potential application of the MOF for the quantitative detection of troponin over a wide concentration range, with the limits of detection and quantification estimated to be 10 fg mL^{-1} and 30 fg mL^{-1}, respectively.

The electrochemical properties of inorganic and organometallic polymers also allow the biosensing of biological analytes. Electrochemically active organometallic polymers enhance electron transfer processes in various amperometric biosensors.[27] The basic principle of an amperometric glucose sensor is enzyme (glucose oxidase) catalyzed oxidation of β-D-glucose by molecular oxygen (O_2) to gluconic acid ($C_6H_{12}O_7$) and hydrogen peroxide (H_2O_2). A successful amperometric glucose sensing relies on the oxidised and reduced species diffusing to and fro the enzyme's active site to transfer electrons (Figure 7.8). This requirement necessitates using diffusional mediators such

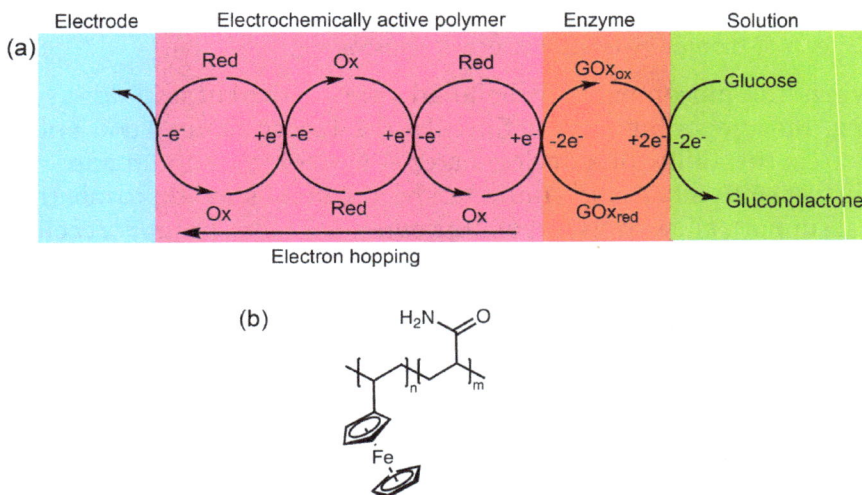

Figure 7.8 (a) Electrochemically active polymer mediated electron transfer for biosensing of glucose. (b) An electrochemically active ferrocene-containing polymer.

as organometallic ferrocene and osmium bipyridine complexes. The need for a diffusional mediator becomes redundant if the enzyme is electrically coupled to the electrode, ensuring that direct electron transfer from the substrate to the enzyme is relayed to an external circuit. Electrochemically active polymers can couple the enzyme to the electrode (Figure 7.8), enhancing the electron transfer process between the substrate, enzyme, and electrodes. A pioneering polymer in this application incorporates an osmium complex into a polyamine to electrically connect a glucose substrate and glucose oxidase to an electrode for glucose sensing.[28] Ferrocene-containing polymers (Figure 7.8) and dendrimers can mediate the electron transfer process and have been used in glucose sensing.[29–31] Ferrocenyl polymers are unique and attractive as redox mediators for several reasons. These reasons include low toxicity, thermal stability, photochemical stability, stable redox states, good enzyme interactions, insensitivity of their redox activity to oxygen and pH perturbations, and electrochemical reversibility at low redox potential. Additionally, it is worth mentioning that ferrocene-containing polymers have been used as redox mediators for the sensing of dopamine,[32] *p*-synephrine,[31] and ascorbic acid.[30,34]

7.3.5 Application as Tissue Engineering Scaffolds

Tissue engineering evolves from the biomaterials field, integrating cell biology and materials engineering principles and techniques. Tissue engineering aims to restore and improve damaged or diseased tissues or organs. Polymers, including inorganic polymers, have been proven to be pivotal scaffolds in tissue engineering. As we have learned from the preceding chapters, the structure of polymers can be more easily controlled than those of metals and ceramics, allowing better control of the properties to meet a desired function. Consequently, compared to ceramics and metals, polymeric

Box 7.2 Criteria of biomaterials used in tissue engineering.

The following three biocompatibility hallmarks are required of scaffolds used in tissue engineering. Biocompatibility, as defined by the United States Food and Drug Administration, is the ability of a device material to perform with an appropriate host response in a specific situation.[36] Importantly, a scaffold should provide the required mechanical support, allow transport of nutrients and oxygen to support cell and tissue growth, and enable the transfer of biochemical signals to modulate cell fate.[37] Therefore, the scaffold must recapitulate the biochemical and mechanical cues of the diseased or damaged tissue to ensure the formation of the desired cell phenotype and tissue. Due to their ability to control their structures through rational synthetic protocols, polymers can be designed to feature these biocompatibility characteristics.

1. *Biodegradability.* Unless the scaffold is not a permanent replacement for diseased or damaged tissue or organs, the scaffolds must be biodegradable or bioresorbable to be acceptable for tissue engineering. Importantly, the by-products of the degradation process must not be toxic or immunogenic to avoid adverse physiological reactions. Also, the by-products must be excretable through normal physiological functions. The degradation rate is equally critical to achieving optimum results. In tissue repair, for instance, if the scaffold degrades before completing its function, the formed tissue may lack support and become defective. On the other hand, if the scaffold persists after completing its function, it may encapsulate the formed tissue, leading to defective tissue, or trigger an immune reaction, culminating in tissue rejection.

2. *Mechanical behavior.* The second guiding principle in designing or selecting a scaffold is to ensure the mechanical properties fit the surrounding tissue. To illustrate, a scaffold designed for bone repair should, in addition to biocompatibility, possess mechanical properties equivalent to those of bone. Suppose the scaffold's hardness (defined as force per unit area of indentation or penetration) is higher than that of bone, the scaffold will likely penetrate the surrounding bones and *vice versa* if the hardness is lower. Further, the mechanical cues of biomaterials (stiffness and deformability) can activate cell surface mechanoreceptors. These mechanical signals play critical roles in cell differentiation, proliferation, and death. Therefore, it is crucial to match the mechanical properties of the scaffold to the mechanobiology of the cell or tissue.

3. *Biochemical compatibility.* Another criterion is that the scaffold must be biochemically compatible with the biological milieu designed to function. This means biochemical interactions between the scaffold and the surrounding cells or tissues should not trigger unwanted reactions. These interactions greatly depend on the chemical composition and morphology of the scaffold. Scaffold surface properties such as surface charge density, surface functional groups, and surface morphology affect cell attachment and signalling. Bulk properties such as solubility (hydrophilicity or lipophilicity), acidity/basicity, and porosity can largely influence cell signalling and migration. Overall, the scaffold surface and bulk properties can affect cell growth, migration, fate, and inflammatory and immunological responses.

scaffolds can be more easily designed and fabricated to meet the criteria required for tissue engineering (Box 7.2) and to achieve the following:

(a) improve naturally occurring processes such as expediting wound healing,
(b) restore dysfunctional cellular processes such as healing diseased tissues,
(c) generate new tissues through cell transplantation, and
(d) attenuate undesirable natural processes such as immune rejection of allografts.

Polysiloxanes, commonly known as silicones, are one of the most widely used inorganic polymers in reconstructive and cosmetic surgery. This widespread use is primarily driven by their biochemical inertness, optical transparency, viscosity, and good oxygen permeability. In addition to using silicone gels as breast implants, testicular prostheses, and pectoral implants, silicone oils are also used as vitreous substitutes after vitrectomy to stabilise and reattach complicated cases of retinal detachment. However, when used as a vitreous substitute, a second surgery may be necessary due to the slow biodegradability of silicones.[35] Although no commercialised applications exist, the properties of polyphosphazenes remain promising for tissue engineering. The properties of these polymers are highly tunable by grafting different side chains or functional molecules onto the polyphosphazene backbone. For example, grafting the antioxidant ferulic acid to a polyphosphazene yields a cross-linkable polymer that is amenable to fabricating a potential free radical-scavenging scaffold for hard tissue engineering. Further, the ability to graft different chemical groups on the polyphosphazene backbone provides a strategy to control cell growth and adhesion in tissue regeneration applications. Replacing imidazolyl pendants with ethyl glycinato groups on the backbone of polyphosphazenes favors increased osteoblast adhesion and growth and increased polymer degradation rate. When fabricated into a 3D scaffold, the poly[(methylphenoxy)(ethyl glycinato) phosphazene] supports a steady osteoblast growth kinetics in contrast to the 2D scaffold, where cell growth declines after 7 days. In summary, the ease of structural modification makes inorganic polymers excellent materials for designing and fabricating scaffolds for tissue engineering.

7.4 Applications in Chemical Synthesis

Most chemical syntheses rely on catalysis. Indeed, catalysis sustains about 85% of the processes in the chemical industry in 2019. From basic physical chemistry, we learned that a catalyst lowers the activation energy and increases the rate of a chemical reaction, making chemical synthesis more sustainable. Chemists and chemical engineers have explored several approaches to improve the efficiency of catalysts. Probably inspired by enzymes, archetypal natural polymeric catalysts, chemists and chemical engineers are designing polymeric catalysts – polymers containing catalytically active sites. One advantage of polymeric catalysts is the opportunity to tune catalyst efficiency by controlling the polymer framework.[38] For example, improved efficiency can be obtained if the polymer framework provides a hydrophobic microenvironment that protects the catalytic sites from deactivation by oxygen or water.[38] Indeed, an amphiphilic polycarbonate incorporating a catalytic rhenium complex within its backbone self-assembles into micelles and sequesters the catalyst within the hydrophobic core, preventing catalyst deactivation by water.[39] Presumably, this sequestration enables the rhenium-containing polycarbonate to catalyze the reduction of carbon(IV) oxide to carbon(II) oxide while suppressing hydrogen generation in an aqueous medium better than the corresponding molecular rhenium complex.[39]

Another hypothesis is that, compared with molecular catalysts, polymeric catalysts are more efficient due to the increased loading of catalytic sites within the polymer framework. For example, amine-functionalised polysiloxanes are more efficient than a molecular amine, ethane-1,2-diamine, in catalyzing the Gewald reaction. Specifically, amine-functionalised polysiloxanes catalyze the Gewald reaction of cyclohexanone, sulfur, and ethyl 2-cyanoacetate in 12 hours, while ethane-1,2-diamine affords the same reaction in 24 hours (Table 7.2).[40] Also, the amount of the rhenium–bipyridine complex in a rhenium-containing polyhedral oligomeric silsesquioxane-based organometallic polymer (POMP) plays a critical role in the catalytic activity of the polymer to convert carbon(IV) oxide to syngas. Specifically, the amount of the rhenium–bipyridine complex tunes the surface area, CO_2 adsorption ability, visible light harvesting efficiency, and photoinduced electron–hole separation efficiency of the POMP, influencing the catalytic properties.[41]

Another factor that informs the development of polymeric catalysts is the possibility of reusability and improved stability. For water-sensitive organometallic catalysts, the polymer framework can stabilise the catalyst by sequestering the catalyst within a hydrophobic core, limiting the interaction with deactivating molecules such as water. Also, catalysts incorporated within insoluble polymers can readily be separated from reaction solutions and reused without activity loss. The cobalt-containing coordination polymer containing a palladium–pyridine catalyst successfully catalyzes the reduction of nitroarenes to the corresponding amines in four consecutive reactions without a significant decrease in catalytic activity (Box 7.3).[42]

Table 7.2. A polymeric amino catalyst is more efficient than a molecular amine catalyst in the Gewald reaction. Adapted from ref. 40 with permission from the Royal Society of Chemistry.

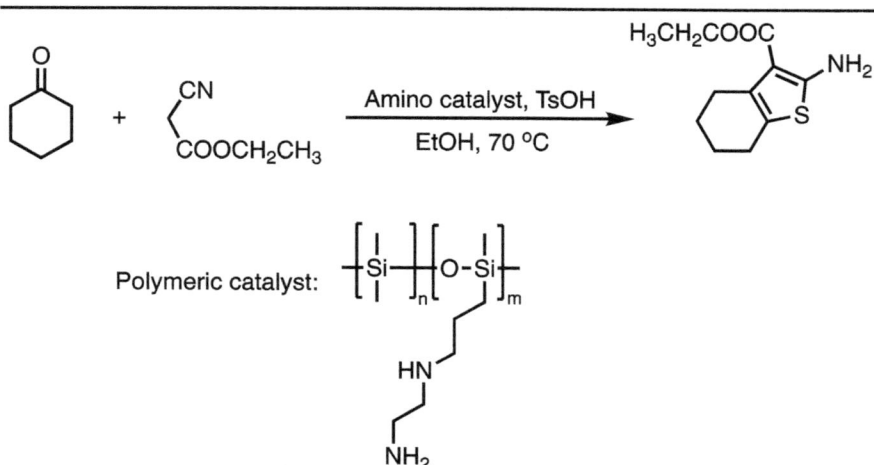

Entry	Catalyst	Time (h)	Yield (%)
1	$NH_2(CH_2)_2NH_2$	24	35
2	Polymeric catalyst	12	58

Box 7.3 Practice question 1.

Question

Metal–organic frameworks (see Chapter 6) are known for their porosity and large surface area. Considering these two mentioned properties, propose an application that we did not discuss in this chapter for metal–organic frameworks.

Answer

The porosity and large surface area of metal–organic frameworks are ideally suited for gas capture and storage. The pores can be functionalised to improve for selective capture and storage of gases. Indeed, literature reports abound demonstrating the use of metal–organic frameworks in the capture and storage of various gases including carbon(IV) oxide.

7.5   Applications in Light Harvesting

On our planet, sunlight is the most ubiquitous and sustainable form of energy, with far-reaching implications for supporting life. Natural plants and photosynthetic bacteria capture and convert sunlight to chemical energy *via* photosynthesis. Inspired by nature, scientists are developing platform technologies capable of harvesting sunlight and transforming it into other forms of energy. An underlying principle of these technologies is energy transfer between an excited state donor fluorophore and a ground state acceptor *via* Forster resonance energy transfer. This energy transfer can be fine-tuned by

(a) molecular level organisation of the donor in a specific orientation to harvest light energy,
(b) optimising the distance between the acceptor and donor and
(c) overlapping the emission spectrum of the donor and the absorption spectrum of the acceptor.

Polymers provide a macromolecular framework to fine-tune the energy transfer process to optimise the photophysics of the excited state of the donor fluorophore and prevent quenching. Multiple donor chromophores can be appended on the polymer backbone for efficient light harvesting and energy transfer processes. Further, donor and acceptor molecules can be grafted and rationally organised on the polymer framework to optimise energy transfer. For example, an efficient energy transfer from the donor to the acceptor is attainable in the polystyrene-*graft*-donor-*co*-polystyrene-*graft*-acceptor copolymer (the donor is $[C_6H_4–CH_2–C_2HN_3–C_6H_4–C\equiv C–Pt(PBu_3)_2–C\equiv C–C_6H_4–C\equiv C–C_5H_6]$; the acceptor is $[C_6H_4–CH_2–C_2HN_3–C_6H_4–C\equiv C–Pt(PBu_3)_2–C\equiv C–C_6H_4–C\equiv C–C_{19}H_9]$).[43] The donor-to-acceptor ratio in the polystyrene backbone is critical to energy transfer efficiency, with a 20 : 1 ratio affording over 95% efficiency. Time-resolved phosphorescence combined with Monte Carlo exciton dynamics simulations indicated an energy transfer mechanism that involves exciton hopping between fluorophores on the polystyrene backbone.

Polymers provide a framework within one platform to harvest sunlight using multiple donor fluorophores. Dendrimers, a highly ordered and branched class of polymers, offer an ideal framework to append different donor and acceptor fluorophores in one platform. An example is a platinum-containing rotaxane dendrimer having a pyrene or zinc(II) porphyrin core as the energy acceptor and anthracene or tetraphenylethene moieties on the periphery as the light-harvesting antenna and energy donor.[44,45] Metalloporphyrins are excellent coordination compounds used in the design of dendrimeric light-harvesting systems. The coordination metal ion in the porphyrin tunes the absorption and emission spectra, allowing the use of different metalloporphyrins as donors or acceptors. A dendrimer designed from a copper(II) porphyrin core and a zinc(II) porphyrin periphery demonstrated this concept because the zinc(II) porphyrin functions as the light-harvesting antenna and donor, while the copper(II) porphyrin acts as the acceptor.[46] Further, transition metal-containing co-ordination polymers are another class of polymeric systems explored as potential light-harvesting systems. The typical metal-to-ligand charge transfer properties of transition metal complexes allow light absorption. Also, d→f energy transfer can sensitise an inner transition metal in a complex of a transition metal–inner transition metal dyad. A coordination polymer incorporating iridium and lanthanide complexes excellently demonstrates this principle. The iridium complex in the dyad harvests light *via* triplet metal-to-ligand charge transfer to sensitise the lanthanide near-infrared luminescence *via* d→f energy transfer from the iridium to the lanthanide ion.[47] Currently, no inorganic or organometallic polymer light-harvesting systems are commercially available, but their potential has been demonstrated in various laboratories.

Review Questions

1. List the properties of polysiloxanes that make them well-suited for drug delivery.
2. Design a polymeric drug that can act through multiple mechanisms to kill cancer cells.
3. Design a redox active polymer capable of redox-triggered self-assembly and disassembly.
4. Explain why a ferrocene-containing polymer may not be suitable for fabricating scaffolds for tissue engineering.
5. Explain why a zinc-containing polymer is not suitable for designing an electro-chemical biosensor.

Further Reading

1. A. S. Abd-El-Aziz, C. Agatemor and W.-Y. Wong, *Macromolecules Incorporating Transition Metals Tackling Global Challenges*, Royal Society of Chemistry, United Kingdom, 2018.
2. J. E. Mark, H. R. Allcock and E. West, *Inorganic Polymers*, Prentice Hall, Englewood NJ, 1992.
3. E. Het-Hawkins and M. Hissler, *Smart Inorganic Polymers. Synthesis, Properties, and Emerging Applications in Materials and Life Sciences*, Wiley-VCH, 2019.

References

1. E. Gianasi, M. Wasil, E. Evagorou, A. Keddle, G. Wilson and R. Duncan, *Eur. J. Cancer*, 1999, **35**, 994–1002.
2. P. Sood, K. B. Thurmond, J. E. Jacob, L. K. Waller, G. O. Silva, D. R. Stewart and D. P. Nowotnik, *Bioconjugate Chem.*, 2006, **17**, 1270–1279.
3. R. Duncan, *Nat. Rev. Drug Discovery*, 2003, **2**, 347–360.
4. X. Zeng, Y. Wang, J. Han, W. Sun, H. Butt, X. Liang and S. Wu, *Adv. Mater.*, 2020, **32**, 2004766.
5. K. Lu, C. He, N. Guo, C. Chan, K. Ni, R. R. Weichselbaum and W. Lin, *J. Am. Chem. Soc.*, 2016, **138**, 12502–12510.
6. K. Lu, C. He, N. Guo, C. Chan, K. Ni, G. Lan, H. Tang, C. Pelizzari, Y.-X. Fu and M. T. Spiotto, *Nat. Biomed. Eng.*, 2018, **2**, 600–610.
7. Y. Shao, B. Liu, Z. Di, G. Zhang, L.-D. Sun, L. Li and C.-H. Yan, *J. Am. Chem. Soc.*, 2020, **142**, 3939–3946.
8. A. Frei, A. D. Verderosa, A. G. Elliott, J. Zuegg and M. A. Blaskovich, *Nat. Rev. Chem.*, 2023, 1–23.
9. A. S. Abd-El-Aziz, C. Agatemor, N. Etkin, D. P. Overy, M. Lanteigne, K. McQuillan and R. G. Kerr, *Biomacromolecules*, 2015, **16**, 3694–3703.
10. A. S. Abd-El-Aziz, C. Agatemor, N. Etkin, D. P. Overy and R. G. Kerr, *RSC Adv.*, 2015, **5**, 86421–86427.
11. A. S. Abd-El-Aziz, C. Agatemor, N. Etkin, R. Bissessur, D. Overy, M. Lanteigne, K. McQuillan and R. G. Kerr, *Macromol. Biosci.*, 2017, **17**, 1700020.
12. R. Namivandi-Zangeneh, E. H. Wong and C. Boyer, *ACS Infect. Dis.*, 2021, **7**, 215–253.
13. S. Bouson, A. Krittayavathananon, N. Phattharasupakun, P. Siwayaprahm and M. Sawangphruk, *R. Soc. Open Sci.*, 2017, **4**, 170654.
14. Z. Yuan, J. Wang, Y. Wang, Y. Zhong, X. Zhang, L. Li, J. Wang, S. F. Lincoln and X. Guo, *Macromolecules*, 2019, **52**, 1400–1407.
15. C. Zheng, L. Qiu, X. Yao and K. Zhu, *Int. J. Pharm.*, 2009, **373**, 133–140.
16. J. Xu, X. Zhu and L. Qiu, *Int. J. Pharm.*, 2016, **498**, 70–81.
17. Y. Peng, X. Zhu and L. Qiu, *Biomaterials*, 2016, **106**, 1–12.
18. C. Chun, S. M. Lee, C. W. Kim, K.-Y. Hong, S. Y. Kim, H. K. Yang and S.-C. Song, *Biomaterials*, 2009, **30**, 4752–4762.
19. A. K. Maparu, P. Singh, B. Rai, A. Sharma and S. Sivakumar, *Mater. Sci. Eng., C*, 2021, **119**, 111577.
20. L. Bai, H. Yan, L. Guo, M. He, T. Bai and P. Yang, *Macromol. Chem. Phys.*, 2021, **222**, 2100283.
21. L. Bai, H. Yan, T. Bai, Y. Feng, Y. Zhao, Y. Ji, W. Feng, T. Lu and Y. Nie, *Biomacromolecules*, 2019, **20**, 4230–4240.
22. B. Pfleiderer, A. Moore, E. Tokareva, J. L. Ackerman and L. Garrido, *Biomaterials*, 1999, **20**, 561–571.
23. P. Rościszewski, J. Łukasiak, A. Dorosz, J. Galiński and M. Szponar, *Macromol. Symp.*, 1998, **130**, 337–346.
24. J. Cao, Y. Zuo, H. Lu, Y. Yang and S. Feng, *J. Photochem. Photobiol., A*, 2018, **350**, 152–163.
25. H. Lu, Z. Hu and S. Feng, *Chem. – Asian J.*, 2017, **12**, 1213–1217.
26. A. S. Basaleh and S. M. Sheta, *J. Inorg. Organomet. Polym. Mater.*, 2021, **31**, 1726–1737.
27. R. Gracia and D. Mecerreyes, *Polym. Chem.*, 2013, **4**, 2206–2214.
28. A. Heller, *J. Phys. Chem.*, 1992, **96**, 3579–3587.
29. M. Şenel, C. Nergiz and E. Çevik, *Sens. Actuators, B*, 2013, **176**, 299–306.
30. M. Cirelli, J. Hao, T. C. Bor, J. Duvigneau, N. Benson, R. Akkerman, M. A. Hempenius and G. J. Vancso, *ACS Appl. Mater. Interfaces*, 2019, **11**, 37060–37068.
31. D. Estrada-Osorio, R. A. Escalona-Villalpando, A. Gutiérrez, L. Arriaga and J. Ledesma-García, *Bioelectrochemistry*, 2022, **146**, 108147.
32. E. Scavetta, R. Mazzoni, F. Mariani, R. Margutta, A. Bonfiglio, M. Demelas, S. Fiorilli, M. Marzocchi and B. Fraboni, *J. Mater. Chem. B*, 2014, **2**, 2861–2867.
33. P. Lach, M. Cieplak, K. R. Noworyta, P. Pieta, W. Lisowski, J. Kalecki, R. Chitta, F. D'Souza, W. Kutner and P. S. Sharma, *Sens. Actuators, B*, 2021, **344**, 130276.
34. I. Pandey and S. S. Jha, *Electrochim. Acta*, 2015, **182**, 917–928.
35. S. Schnichels, N. Schneider, C. Hohenadl, J. Hurst, A. Schatz, K. Januschowski and M. S. Spitzer, *PLoS One*, 2017, **12**, e0172895.
36. United State Food and Drug Administration, *Use of International Standard ISO 10993-1, Biological Evaluation of Medical Devices – Part 1: Evaluation and Testing within a Risk Management Process*, U.S. Department of Health and Human Services, Food and Drug Administration, Center for Device and Radiological Health, Center for Biologics Evaluation and Research. September 4, 2020.
37. M. I. Echeverria Molina, K. G. Malollari and K. Komvopoulos, *Front. Bioeng. Biotechnol.*, 2021, **9**, 617141.

38. A. Deratani, G. D. Darling, D. Horak and J. M. Frechet, *Macromolecules*, 1987, **20**, 767–772.
39. F. Ren, K. Chen, L. Qiu, J. Chen, D. J. Darensbourg and L. He, *Angew. Chem.*, 2022, **134**, e202200751.
40. Z.-J. Zheng, L.-X. Liu, G. Gao, H. Dong, J.-X. Jiang, G.-Q. Lai and L.-W. Xu, *RSC Adv.*, 2012, **2**, 2895–2901.
41. W.-J. Wang, K.-H. Chen, Z.-W. Yang, B.-W. Peng and L.-N. He, *J. Mater. Chem. A*, 2021, **9**, 16699–16705.
42. S. Aghajani and M. Mohammadikish, *Langmuir*, 2022, **38**, 8686–8695.
43. Z. Chen, H.-Y. Hsu, M. Arca and K. S. Schanze, *J. Phys. Chem. B*, 2015, **119**, 7198–7209.
44. W.-J. Li, X.-Q. Wang, W. Wang, Z. Hu, Y. Ke, H. Jiang, C. He, X. Wang, Y.-X. Hu and P.-P. Jia, *Giant*, 2020, **2**, 100020.
45. W.-J. Li, H. Jiang, X.-Q. Wang, D.-Y. Zhang, Y. Zhu, Y. Ke, W. Wang and H.-B. Yang, *Mater. Today Chem.*, 2022, **24**, 100874.
46. H. Lee, Y.-H. Jeong, J.-H. Kim, I. Kim, E. Lee and W.-D. Jang, *J. Am. Chem. Soc.*, 2015, **137**, 12394–12399.
47. L. Li, S. Zhang, L. Xu, Z.-N. Chen and J. Luo, *J. Mater. Chem. C*, 2014, **2**, 1698–1703.

Subject Index

www.ingramcontent.com/pod-product-compliance
Lightning Source LLC
Chambersburg PA
CBHW061928190326
41458CB00009B/2692